Ana Beatriz Alves de Araújo
Erika S. Alves Graciano de Vasconcelos

Agricultural Sciences

AF319430

Ana Beatriz Alves de Araújo
Erika S. Alves Graciano de Vasconcelos

Agricultural Sciences

Technologies and Sustainability

ScienciaScripts

Imprint

Any brand names and product names mentioned in this book are subject to trademark, brand or patent protection and are trademarks or registered trademarks of their respective holders. The use of brand names, product names, common names, trade names, product descriptions etc. even without a particular marking in this work is in no way to be construed to mean that such names may be regarded as unrestricted in respect of trademark and brand protection legislation and could thus be used by anyone.

Cover image: www.ingimage.com

This book is a translation from the original published under ISBN 978-620-6-76220-1.

Publisher:
Sciencia Scripts
is a trademark of
Dodo Books Indian Ocean Ltd. and OmniScriptum S.R.L publishing group

120 High Road, East Finchley, London, N2 9ED, United Kingdom
Str. Armeneasca 28/1, office 1, Chisinau MD-2012, Republic of Moldova, Europe
Managing Directors: Ieva Konstantinova, Victoria Ursu
info@omniscriptum.com

Printed at: see last page
ISBN: 978-620-3-23042-0

Copyright © Ana Beatriz Alves de Araújo, Erika S. Alves Graciano de Vasconcelos
Copyright © 2025 Dodo Books Indian Ocean Ltd. and OmniScriptum S.R.L publishing group

CHAPTER 1

USE OF REMOTE SENSING THROUGH VEGETATION INDICES TO CHARACTERIZE VEGETATION COVER IN THE BRAZILIAN SEMIARID REGION

Ana Beatriz Alves de ARAÚJO;

Agricultural and Environmental Engineer, PhD, UFERSA, Mossoró/RN

Suedêmio Lima e SILVA;

Agricultural Engineer, PhD, UFERSA, Mossoró/RN.

INTRODUCTION

Geoprocessing techniques have been widely used to monitor biophysical characteristics and anthropogenic actions on Earth, so that in recent years land use zoning has been carried out in order to identify areas that need appropriate management, both in relation to soil erosion and vegetation vigor, helping to establish sustainable use conditions (OLIVEIRA et al., 2012).

Currently, the use of spectral indices derived from remote sensing has aroused the interest of professionals looking for ways to measure biomass and vegetation cover; since quantifying biomass and also the amount of vegetation in agricultural areas from soil analysis can take time to generate a result and have a high cost, especially in semi-arid regions, where productivity is generally low (EISFELDER et. al, 2012).

It is understood that one of the most important indices for assessing the condition of an ecosystem is vegetation cover, and its condition is indirectly linked to anthropogenic activities, farming, deforestation and desertification (WANG et al., 2015). However, effectively quantifying the changes that occur in vegetation over a long period of time is still a challenge for agricultural and environmental studies. And one way that has proven efficient for this measurement is the Normalized Difference Vegetation Index (NDVI), which is derived from the difference between the relative reflectance in the red band and the near infrared band (ZHENG et al.; 2018).

The Normalized Difference Vegetation Index (NDVI) is a widely used index. However, in regions with sparse vegetation or soils that generate high reflectance values, for example dry soils; they can produce biased results with green biomass and vegetation cover (HORNING et al., 2010; YAGCI et al., 2014; LEE et al., 2016). To reduce variability caused by ground background reflectance, illumination, and view angle variation, remote sensing scientists have addressed this through the use of new spectral vegetation indices that can accommodate greater variability due to ground reflectance, for example; Soil Adjusted Vegetation Index (SAVI). NDVI continues to be used to estimate biomass and cover in areas with varied vegetation types, but its use in semi-arid areas is becoming increasingly scarce, especially in regions with sandy soils (ZHENG, et. al. 2018).

In this way, vegetation indices provide information for understanding environmental phenomena, as well as the possibility of strategic planning in different situations, such as urban planning (ALVARENGA; MORAES, 2014).

Thus, the research presented here is justified in demonstrating the use of vegetation indices through remote sensing techniques, namely the Normalized Difference Vegetation Index (NDVI) and the Soil Adjusted Vegetation Index (SAVI), highlighting the importance of selectivity in choosing the most appropriate vegetation index for the area and for the purpose of the proposed research, which is to represent and discriminate the vegetation cover of the municipality of Mossoró/RN.

LITERATURE REVIEW

Geoprocessing and Vegetation Indices

The use of geotechnologies for remote sensing is quite numerous, and some areas stand out in terms of the application of sensing techniques: forestry and the environment; water resources; geology; urban planning; cartography; etc.

By successively scanning the planet's surface, satellite images make it possible to study and monitor anthropogenic natural phenomena such as deforestation. Various natural phenomena, such as flooding and soil erosion, are intensified or aggravated by human action. According to Florenzano (2011), with the use of satellite images, it is possible to identify, calculate and monitor the growth of different types of area: deforested, affected by fire, subjected to erosion processes, etc., stating that the multi-temporal aspect of satellite images makes it possible to evaluate and monitor deforested areas. Based on the interpretation of digital images, maps of deforested areas from different dates can be generated, allowing the interpretation of various aspects inherent to the area studied.

According to Tôsto et al. (2014), geoprocessing involves the use of computer tools to process and analyze geographic data. The set of these tools, connected in Geographic Information Systems (GIS), allows for the analysis and cross-referencing of data from various sources, promoting the extraction of information and decision-making. A clear difference between GIS and other conventional systems is the ability to carry out a spatial analysis of data, since it has a set of associated information in its files. A GIS is suitable for representing a huge variety of spatial data, such as the location and delimitation of areas of interest, topography, distribution networks, among other attributes. One of the main advantages of using Geographic Information Systems is the range of software available on the market, many of which is free.

Normalized Difference Vegetation Index (NDVI)

The normalized difference vegetation index (NDVI) is an index calculated from the reflectance values of the near infrared (ρNIR) and visible red (ρRED) bands.

NDVI is an index related to the conditions and quantities of vegetation (BORATTO; GOMIDE, 2013). It is sensitive to chlorophyll and other pigments that capture solar radiation (RISSO et al., 2009).

According to Silva et. al (2010), the NDVI is a useful IV for determining the biophysical parameters of vegetation and, by considering the ratio in its calculation, there is a reduction in multiplicative noise such as lighting, cloud shadows, atmospheric factors and others. Jensen (2009) describes some important factors that should be considered when applying this index, taking into account seasonal and inter-annual changes in the study, as well as the monitoring of vegetation cover. The NDVI results in a monochromatic spectral image whose maximum and minimum values vary between 1 and -1, respectively. Values closer to 1 indicate regions with a low concentration of gray levels and represent areas with vegetation. While values close to -1 show darker shades of gray levels and describe the presence of water resources, exposed soil and other classes.

According to Melos and Sales (2011), NDVI is the most widely used and best-known vegetation index, as it allows the density and state of vegetation on the surface to be monitored, enabling its spatialization and evolution over time to be characterized.

According to Huete & Tucker (1991), NDVI values for exposed soils are normally between 0.05 and 0.30, and due to the great variability of the soil's optical properties, it is not possible to define a range of strict NDVI values for soils with little or no vegetation. In these cases, the use of other indices is recommended, such as SAVI (soil-adjusted vegetation index), one of the objects of this research.

Soil Adjusted Vegetation Index (SAVI)

The soil-adjusted vegetation index was developed with the aim of mitigating the effects of the soil when processing digital images. According to Abreu (2018), the purpose of the index is to measure vegetation activity and its seasonal variation. According to Viganó, Borges and França-Rocha (2011), SAVI is the index that best portrays soil variability in an environment that features the Caatinga biome, since it has a soil adjustment constant in its equation, responsible for highlighting color variations in the product, facilitating geospatial analysis.

According to Leite et al. (2017), the adjustment factor varies according to the density of the local vegetation. For lower density vegetation cover, the L factor is equivalent to 1.0; for areas with intermediate vegetation cover, the factor will be 0.5; for high densities, the factor is equivalent to 0.25. The SAVI index will correspond to the NDVI index when the L factor is equal to zero (DEMARCHI et al., 2011).

Both indices are indicators of vegetation density and condition. SAVI is nothing more than an adaptation of NDVI. It uses the L adjustment factor in order to minimize the effect of the presence of exposed soil. According to Huete (1988), the adjustment factor allows for the removal of the effect of light or dark soils, thus mitigating the effects of soil background.

In the soil-adjusted vegetation index, positive values indicate areas with vegetation, while negative values represent areas without vegetation, clouds and/or bodies of water (ALVARENGA; MORAES, 2014).

Leite et al. (2017), in a study on the temporal analysis of the NDVI and SAVI vegetation indices at the Itaitinga experimental station, used digital images from the Landsat 8 sensore OLI satellite. Corroborating the characteristics of the area, their results indicated little anthropogenic alteration. Both indices, NDVI and SAVI, obtained positive results, as this is an area with a high level of vegetation, but there was a greater suitability for the use of NDVI, characterized by the high density of local vegetation.

Viganó, Borges and França-Rocha (2011), in a study on the performance of the SAVI and NDVI vegetation indices based on digital image analysis with the aim of identifying which index best represented and discriminated between the vegetation cover of the study area, chose SAVI as the best index for the object, since a greater number of classes of values corresponding to the vegetated area were presented; however, they stress the importance of considering the research area.

MATERIAL AND METHODS

The work was carried out in the municipality of Mossoró, which is located in the state of Rio Grande do Norte, in the Northeast region of Brazil, with Longitude W 37° 20' 00", Latitude S 5° 11" 00", and an average altitude of 18 m. It has a territorial unit area of 2,099,333 km(2), with a population density of 123.9 inhab./km²). It has a territorial unit area of 2,099.333 km², with a population density of 123.76 inhabitants/km², corresponding to 3.9% of the state's area (IBGE, 2010).

The predominant climate is semi-arid and, according to the KÖPPEN classification, of the BSw'h' type, i.e. dry, very hot, with the rainy season concentrated between summer and fall, and a dry season of 7 to 9 months, with irregular rainfall. It has an average temperature of 27.4°C and relative humidity of 70% (IDEMA, 2000).

Among the soils present in the municipality, Cambissolos predominate, highly fertile soils with a clayey texture, generally shallow and moderately drained, occupying areas of flat relief with a predominance of hyperxerophilic caatinga, whose plants are well adapted to local desiccation (EMBRAPA, 2006).

Satellite images from the Operational Land Imager (OLI) sensor of the Landsat 8 satellite were used to cover the entire area of the municipality of Mossoró, obtained from the UFGS (Science for a Changing World) database available at (https://earthexplorer.usgs.gov/). The images correspond to the orbit with UTM cartographic projection, WGS 84 datum and Zone 24, with dates of passage on May 15, June 16, July 18, August 3 and September 20, all dates referring to the year 2016. The landsat 8 satellite was chosen because it has a satisfactory pass frequency of 16 days, as well as images with a resolution of 30 meters per pixel with 12 bits per pixel, providing a large amount of information per pixel. Five categories were used; these categories are divided into different placements (C1 red color, C2 blue color, C3 yellow color, C4 light green color and C5 dark green color).

QGis 2.14.16 software was used for the calculations and image formation. Excel® was also used for part of the analysis. There was no need to correct the data for possible noise caused by shadows or clouds, as the images used were obtained with only 10% or less clouds.

Images from the Landsat 8 satellite were used. These images refer to bands 4 and 5 of the satellite, bands with spectral resolutions of red (0.630 - 0.680 µm) and near infrared (0.845 - 0.885 µm).

The vegetation indices studied were NDVI and SAVI. These are the spectral vegetation indices most commonly used to estimate vegetation by satellites. With the reflectance data it is possible to obtain the NDVI using the following equation (proposed by Rouse et al., 1973).

$$NDVI = \frac{\rho NIR - \rho Red}{\rho NIR + \rho Red} \tag{1}$$

Where ρNIR refers to the near infrared band and ρRed to the red band on the Landsat 8 sensor. The NDVI is an index related to the condition and quantities of vegetation (BORATTO; GOMIDE, 2013), it is sensitive to chlorophyll and other pigments that capture solar radiation (RISSO et al., 2009), and the positive point of its use is the better visualization of the real health conditions of the plants present in the sampling area.

To mitigate these effects from the soil, the soil-adjusted vegetation index (SAVI) was created, which can be observed with the following equation.

$$SAVI = \frac{(\rho NIR - \rho Red)}{L + \rho NIR + \rho Red} \cdot (1 + L) \tag{2}$$

Where ρNIR is the radiant flux in the near infrared, and ρRED is the radiant flux in the visible red region, represented by bands 4 and 3 on the Landsat 4-5 Sensor TM satellite and bands 5 and 4 on the Landsat 8 Sensor OLI/TIRS, respectively. L represents the adjustment factor for correcting the effect of ground glare, which varies according to the density of vegetation cover. The adjustment factor varies with vegetation density. For lower density vegetation cover, the L factor is 1.0; for intermediate density vegetation cover, the factor is 0.5, and for high densities, 0.25 (HUETE, 1988). These values in the L factor present in SAVI allow for a better distinction of the condition of vegetation health, taking into account each type of vegetation that is present within the sampled image, thus reducing the effects of the presence of soil in the radiation spectrum when using SAVI, which does not occur with NDVI. The SAVI index is equivalent to the NDVI index when the L factor is equal to zero (DEMARCHI et al., 2011).

RESULTS AND DISCUSSION

By applying the equations to each band suggested for the calculations, NDVI data was obtained for the municipality of Mossoró/RN.

The values for clouds and water were negative in both indices, a similar result found by Santiago et al. (2009) when analyzing vegetation cover using the Vegetation Indices (NDVI, SAVI and IAF) around the Botafogo Dam in the state of Pernambuco.

Areas with lighter colors refer to a denser type of vegetation, while darker colors refer to areas of exposed soil or incipient vegetation. The highest NDVI values correspond to the highest Digital Numbers (DN), which are related to areas of vegetation with greater vigor (BORATTO; GOMIDE, 2013). While the lowest values correspond to low ND, representing areas of stressed vegetation, much less dense or even bare areas (JENSEN, 2009).

According to Ferreira (2013), rainfall can also affect the values of the indices, as it reduces the radiation reflected by the soil and, in the case of the Caatinga, increases the leaf area index of the vegetation after a rainy event.

Where the data found showed a large oscillation in the amount of vegetation in each class, possibly driven by the change in climate during the sampling period, since 36.9 and 28.3 mm of rain were recorded in April and May of this year's study, which may have altered the behavior of the vegetation, triggering a process of leaf growth, which would alter the values of both NDVI and SAVI.

Figure 1 shows the NDVI values for the 5 measurement times, where it can be seen that the images show a similar behavior for all the dates analyzed, with vegetation indices that are close in values within each category.

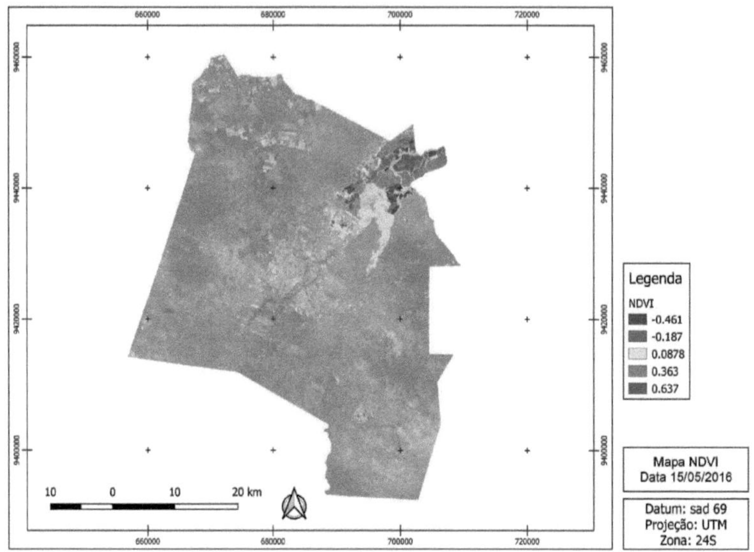

a) NDVI map for 15/05/2016.

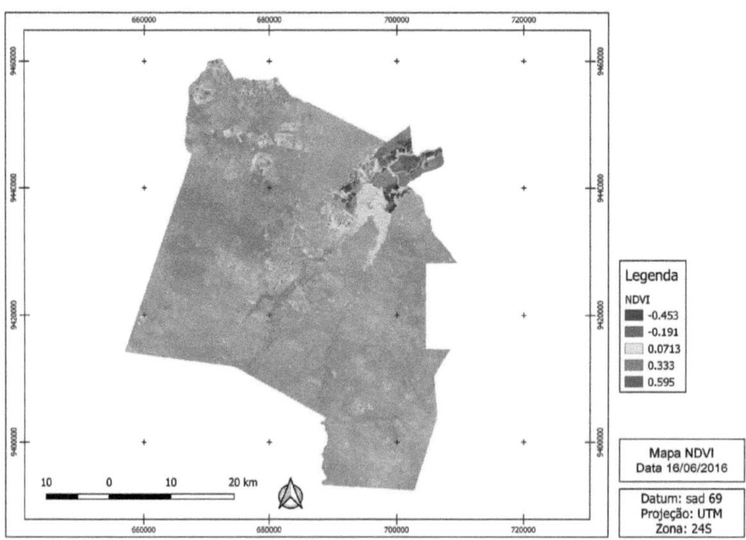

b) NDVI map for 16/06/2016.

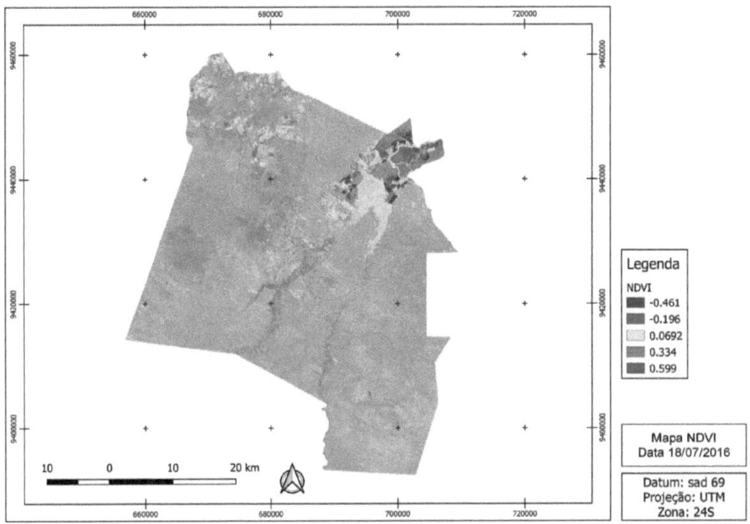

c) NDVI map for 18/07/2016

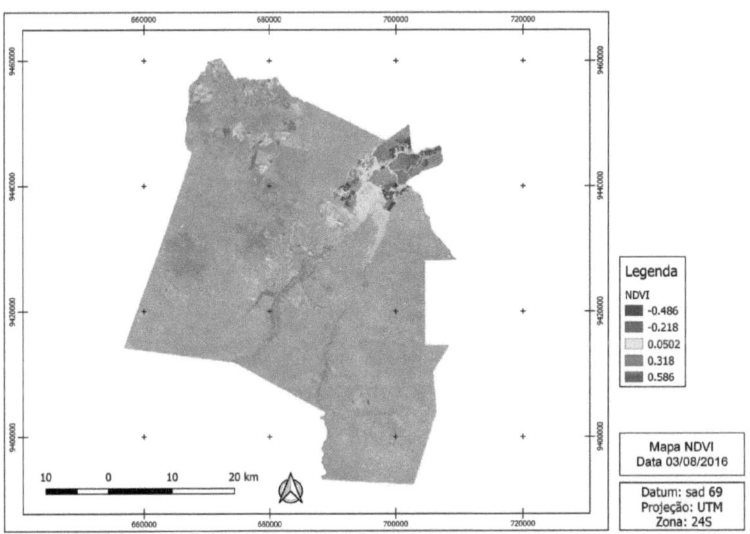

d) NDVI map for 18/07/2016.

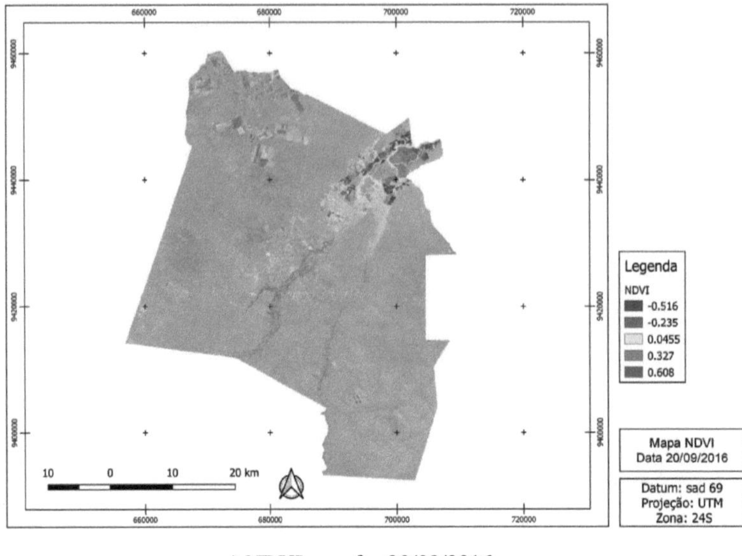

e) NDVI map for 20/09/2016

Figure 1. Unsupervised classification, based on images formed from NDVI values for the period evaluated.

It can be seen in the images above that the locations represented for C5 are more frequent in the images from 15/05/2016 and 16/06/2016, which is in line with the rainy season in the Mossoró/RN region, which takes place between the months of May and June, thus showing greater vegetative vigor during these two periods. The image for 20/09/2016 shows a greater presence of yellow, blue and red colors, suggesting that with the cessation of rainfall in the region, the predominant hyperxerophytic vegetation in the Caatinga begins to undergo a period of water stress, resulting in the senescence of the leaves present in the native vegetation, thus causing a greater emergence of areas with low vegetative vigor.

Figure 2 shows the oscillation of the NDVI values found for the city of Mossoró/RN at five points in 2016, where these values ranged from a minimum of -0.453 on July 18 to a maximum of 0.637 on June 16, as well as showing extreme determination between the values ($R^2 = 0.99$).

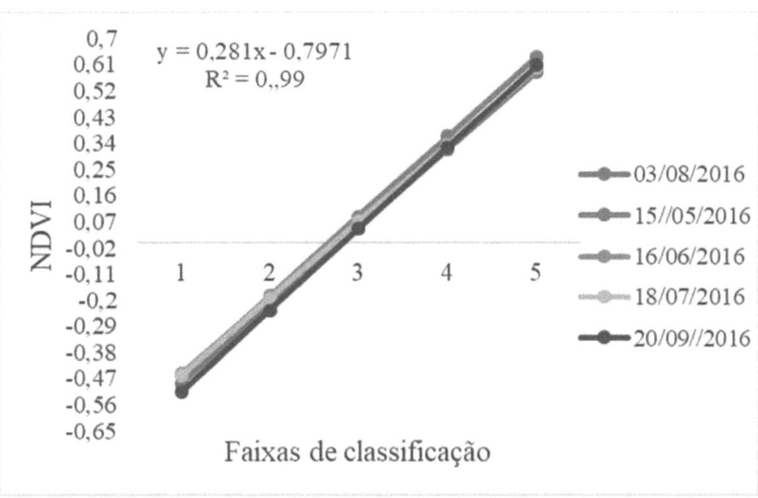

Figure 2. NDVI behavior found for the city of Mossoró/RN on 5 dates in 2016, with variation in NDVI values comparing the 5 collection times.

The SAVI method, on the other hand, with its difference being the adjustment factor to the predominant type of vegetation, has the characteristic of showing higher values within each of the classes compared to the NDVI images. Figure 3 shows the SAVI values for the 5 measurement times. The images show a similar behavior for all times, with vegetation indices that are close in values within each category.

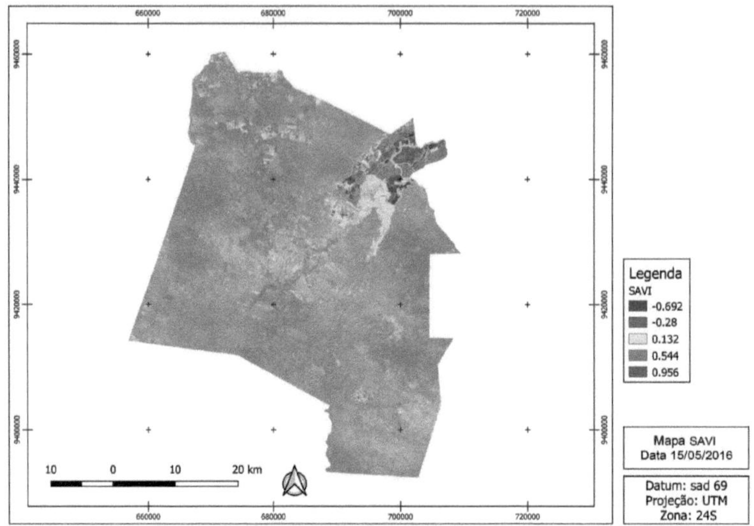

a) SAVI map for 15/05/2016.

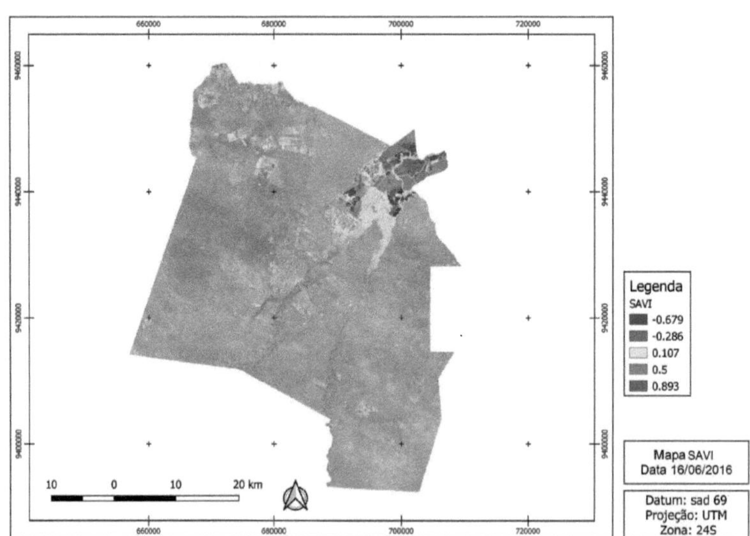

b) SAVI map for 16/06/2016.

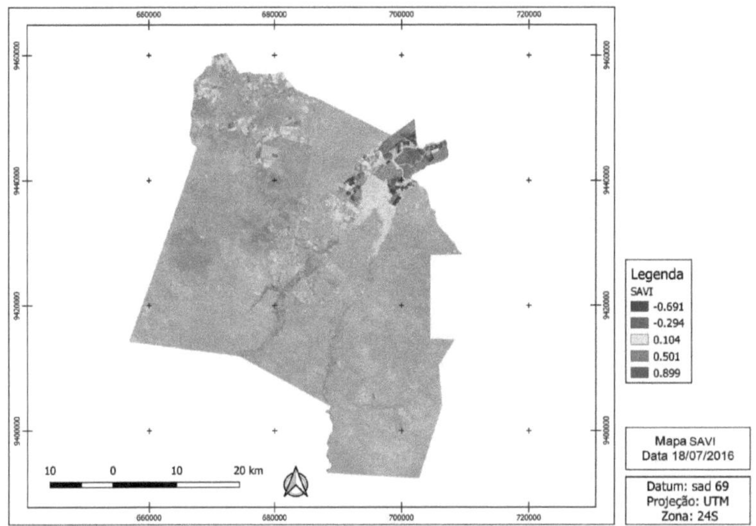

c) SAVI map for 18/07/2016

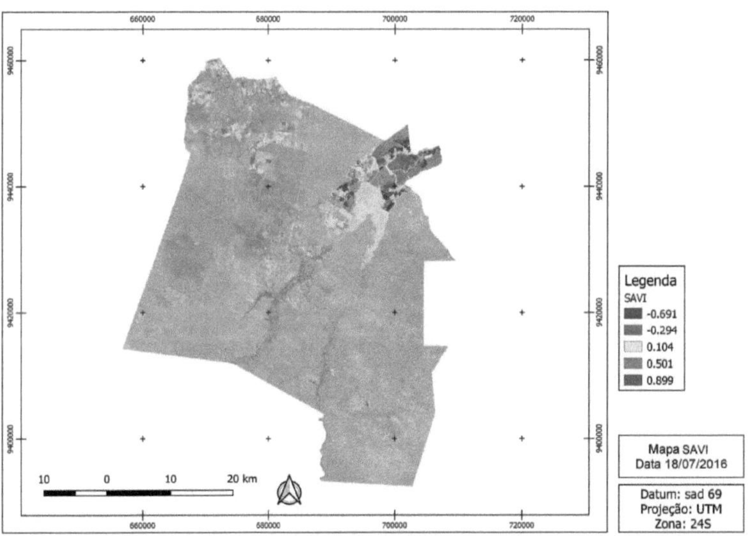

d) SAVI map for 18/07/2016.

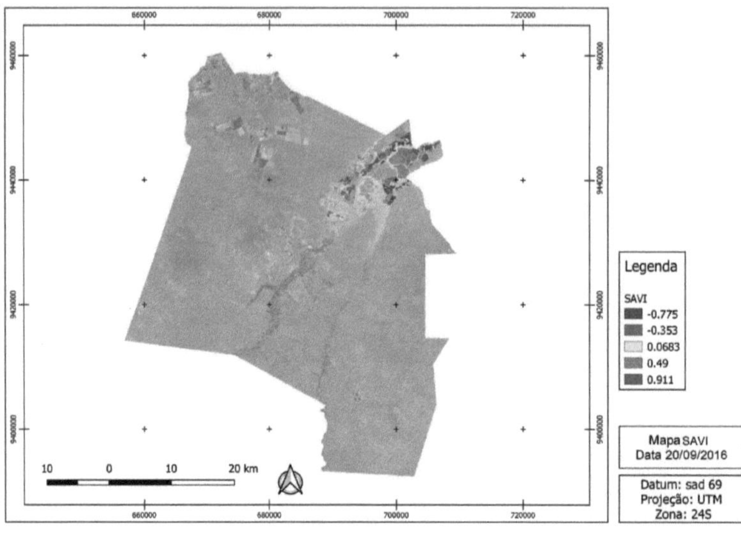

e) SAVI map for 20/09/2016.

Figure 3 - Unsupervised classification, based on images formed from SAVI values for the period evaluated.

It can be seen in the images above that the locations represented for C5 are more frequent in the images from 15/05/2016 and 16/06/2016, just as in the case of NDVI, but you can find higher values in the images than those present in NDVI within each category, thus showing the effect caused by the use of the adjustment factor that is set for SAVI.

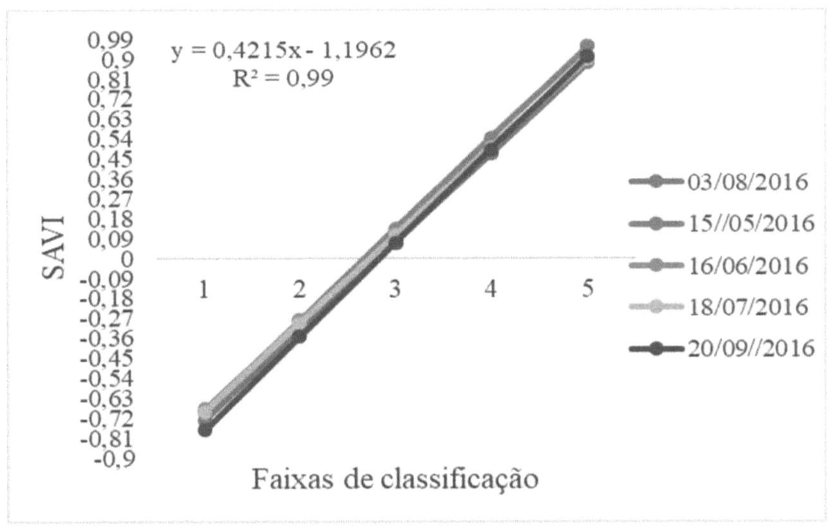

Figure 4 - SAVI behavior found for the city of Mossoró on 5 dates in 2016 - variation in SAVI values comparing the 5 collection times.

Figure X above shows the oscillation of the NDVI values found for the city of Mossoró/RN at five points in 2016. These values ranged from a minimum of -0.775 to a maximum of 0.911, all on September 20, 2016, as well as showing extreme determination between the values ($R^2 = 0.99$).

To avoid biased results, they were divided into five classes for both methods. The index resulting from the calculation and the SAVI and NDVI values found for the 5 evaluation times selected for the work can be seen below in Table 2.

Table 2. SAVI and NDVI values for each sampling date.

Dates	C1	C2	C3	C4	C5
SAVI					
15/05/2016	-0,692	-0,280	0,132	0,544	0,956
16/06/2016	-0,679	-0,286	0,107	0,5	0,893
18/07/2016	-0,691	-0,294	0,104	0,501	0,899
03/08/2016	-0,729	-0,327	0,0753	0,47	0,880
20/09/2016	-0,775	-0,353	0,0683	0,49	0,911
NDVI					
15/05/2016	-0,461	-0,187	0,0878	0,363	0,637
16/06/2016	-0,453	-0,191	0,0713	0,333	0,595
18/07/2016	-0,461	-0,196	0,0692	0,334	0,599
03/08/2016	-0,486	-0,218	0,0502	0,318	0,586
20/09/2016	-0,516	-0,235	0,0455	0,327	0,608

The values for each index show that the SAVI values are higher than the NDVI values at all times. These higher values than the NDVI values were caused by the fact that the SAVI has an adjustment value called the L value, which in this study was assumed to be 0.5.

In the work by Braz et al. (2015), the results for the SAVI indices showed a greater presence in the vegetation indices, because, according to the author, the constant "L" reduces the influence of the soil's spectral response in the SAVI calculation.

This result was also found by Rêgo et al. (2012), where NDVI showed lower indices than SAVI when comparing NDVI and SAVI vegetation indices in the municipality of Cariri/PB. The SAVI indices correspond to targets with a higher density of vegetation.

SAVI in multitemporal analysis results in information about an increase in both the spatialization and density of vegetation in the municipality of Mossoró/RN.

It is worth noting that vegetation indices should not be the only parameter used in studies of changes in the dynamics of vegetation cover, but should be used as an important indicator of these changes, which should be studied in greater detail, especially through fieldwork.

CONCLUSIONS

Oscillating values in terms of the representative areas of the classes showed this behavior, possibly due to the characteristics of the hyperxerophytic vegetation present in the Caatinga environment, whose main feature is a rapid response to changes in the water regime.

Part of this behavior may come from rainfall conditions present at varying times throughout the year, where the rains are concentrated in the first few months of the year and stop almost completely in the last few months.

It was possible to observe the positive use and viability of remote sensing techniques as decision support in monitoring vegetation cover.

The NDVI values were lower than the SAVI values.

Negative values were observed in both indices, which could mean the presence of exposed soil or areas with water.

Most of the red and blue sites are saline areas near the municipality of Assú-RN.

REFERENCES

ALVARENGA, A. S; MORAES, M. F. Digital processing of LANDSAT - 8 images to obtain NDVI and SAVI vegetation indices for characterizing vegetation cover in the municipality of Nova Lima - MG. In: MundoGeo Magazine, 77, July/August, 2014. Available at: http://mundogeo.com/blog/2014/06/10/processamento-digital-de-imagens-landsat-8-para-obtencao-dos-indices-de-vegetacao-ndvi-e-savi-visando-a-caracterizacao-da-cobertura-vegetal-no-municipio-de-nova-lima-mg/Acesso: 12 Dec 2022.

BORATTO, I. M. P.; GOMIDE, R. L. Application of the NDVI, SAVI and IAF vegetation indices to characterize vegetation cover in the northern region of Minas Gerais. In: XVI Brazilian Symposium on Remote Sensing (SBSR), Foz do Iguaçu-PR. Proceedings, São José dos Campos: INPE. p. 7345-7352, 2013.

BRAZ, A. M.; ÁGUAS, T. A.; GARCIA, P. H. M. Analysis of NDVI and SAVI vegetation indices and Leaf Area Index (IAF) for comparison of vegetation cover in the Ribeirãozinho stream watershed, municipality of Selvíria - MS. Revista Percurso - Nemo, v. 7, n.2, p. 05-22, Maringá, 2015.

DEMARCHI, J. C.; PIROLI, E. L.; ZIMBACK, C. R. L. Temporal analysis of land use and comparison between NDVI and SAVI vegetation indices in the municipality of Santa Cruz do Rio Pardo - SP using Landsat - 5 images. Revista Raega, v.21, p.234-271, 2011.

EISFELDER, C., KUENZER, C., DECH, S. Derivation of biomass information for semi-arid areas using remote-sensing data. Int. J. Remote Sens. 33, (2012) 2937-2984.

EMBRAPA (Campinas - Sp). LANDSAT Map - Land Remote Sensing Satellite. 2013. Available at: <https://www.cnpm.embrapa.br/projetos/sat/conteudo/missao_landsat.html>. Accessed on: 10 Dec. 2017.

EMBRAPA. National Soil Research Center. Soil classification system. Brasília: Embrapa Information Production, 2006. 306 p.

FERREIRA, E. M. Geoprocessamento aplicado ao monitoramento ambiental, texto: análise quantitativa de parâmetros biofísicos de bacia hidrográfica obtidos por sensoriamento remoto. Electronic Journal of Civil Engineering. Volume 7, no. 1. 2013.

FLORENZANO, Teresa Gallotti. Introduction to remote sensing / Teresa Gallotti Florenzano. -- 3. ed. expanded and updated. -- São Paulo: Oficina de Textos, 2011.

HORNING, N., ROBINSON, J.A., STERLING, E.J., TURNER, W., SPECTOR, S. Remote Sensing for Ecology and Conservation: A Handbook of Techniques. Oxford University Press. 2010.

HUETE, A.R. A soil-adjusted vegetation index (SAVI). Remote Sensing of Environment, v. 25, p. 295-309, 1988. Available at: <https://www.researchgate.net/publication/220040775_Huete_A_R_A_soiladjusted_vegetation_index_SAVI_Remote_Sensing_of_Environment>. Accessed on: Dec. 15, 2017.

HUETE, A. R.; TUCKER, C. J. Investigation of soil influence in AVHRR red and near infrared vegetation index imagery. International Journal of Remote Sensing, Basingstoke, v.12, p.1223-1242. 1991.

ENVIRONMENTAL DEFENSE INSTITUTE (IDEMA). Profile of your municipality. Mossoró: 2000. Available at: http://www.idema.rn.gov.br. Accessed on: Dec. 15, 2017.

JENSEN, J. R Sensoriamento Remoto do Ambiente: uma perspectiva em recursos terrestres. 2ª ed., São José dos Campos: Parêntese, 2009.

LATORRE, M.; CARVALHO JÚNIOR, O. A.; CARVALHO, A. P. F.; SHIMABUKURO, Y. E. Atmospheric correction: concepts and foundations. Espaço & Geografia, v.5, n.1, p.153-178, 2002.

LEE, M.H., LEE, S.B., EO, Y.D., PYEON, M.W., MOON, K.I., HAN, S.H. Analysis on the effect of Landsat NDVI by atmospheric correction methods. In: Advances in Civil, Architectural, Structural and Constructional Engineering: Proceedings of the International Conference on Civil, Architectural, Structural and Constructional Engineering, Dong-A University, CRC Press, Busan, South Korea, August 21-23, 2015.p. 375.

LEITE, Ana Paula; SANTOS, Glaucia Regina; SANTOS, Jannaylton Éverton Oliveira. Temporal analysis of the ndvi and savi vegetation indices at the itatinga experimental station using landsat 8 images. Revista Brasileira de Energias Renováveis, Curitiba, v. 6, n. 4, p.606-623, apr./may 2017. Available at: Accessed on: 11 Apr. 2019.

LINHARES, S.; GEWANDSZBAJDER, F. Biologia Hoje - Vol 3. São Paulo: Ática, 1998.

LOPES, H. L.; ACCIOLY, L. J. O.; CANDEIAS, A. L. B.; SOBRAL, M. C. Analysis of vegetation indices in the Brígida River basin, Sertão do Estado de Pernambuco. In: III Brazilian Symposium on Geodetic Sciences and Geoinformation Technologies. Recife - PE, p. 01 - 08, jul., 2010.

MELO, Ewerton Torres; SALES, Marta Celina Linhares. Application of the Normalized Difference Vegetation Index (NDVI) to analyze the environmental degradation of the Riacho dos Cavalos watershed, Crateús - CE. 2011. Available at: Accessed on: 11 abr. 2019.
NOVO, E. M. L. M. Remote sensing: principles and applications. 3rd ed., São Paulo: 2008.

OLIVEIRA, T. H.; GALVÍNCIO, J. D; SILVA, J. S.; SILVA, C. A.V.; SANTIAGO, M. M.; MENEZES, J. B.; SILVA, H. A.; PIMENTEL, R. M. M. Evaluation of Vegetation Cover and Albedo of the Moxotó River Basin with Landsat 5 Satellite Images. In: XIV Simpósio Brasileiro de Sensoriamento Remoto, Anais, Natal: INPE, April, p. 2865-2872, 2009.

OLIVEIRA, L. M. T. Study of Brazilian Phytoecological Regions by FAPAR/NDVI and relationships with time series of rainfall data. 2008. 226 p. Thesis (PhD in Civil Engineering) - Federal University of Rio de Janeiro, Rio de Janeiro. 2008.

OLIVEIRA, L. M. M.; MONTENEGRO, S. M. G. L.; ANTONINO, A. C. D.; SILVA, B. B.; MACHADO, C. C. C.; GALVÍNCIO, J. D. Quantitative analysis of watershed biophysical parameters obtained by remote sensing. Revista Pesquisa Agropecuária Brasileira, v. 47, n.9, p. 1209 - 1217, Sep. 2012.

POLONIO, V. D. Vegetation indices for measuring carbon stock in sugarcane areas. 2015. 73p. Dissertation (Master's Degree in Agronomy: Energy in Agriculture). São Paulo State University, Botucatu School of Agronomic Sciences, 2015.

REGO, S. C. A., LIMA, P. P. S., LIMA. M. N. S., MONTEIRO. T. R. R., Comparative analysis of the NDVI and SAVI vegetation indices in the municipality of São Domingos do Cariri-PB. Revista Geonorte, Special Edition, V.2, N.4, p.1217 - 1229, 2012.

RISSO, J.; RIZZI, R.; EPIPHANIO, R. D. V.; RUDORFF, B. F. T.; FORMAGGIO, A. R.; SHIMABUKURO, Y. E.; FERNANDES, S. L. Potentiality of vegetation indices EVI and NDVI from MODIS products in the spectral separability of soybean areas. In: XIV Symposium on Remote Sensing (SBSR), Natal - RN. Proceedings, São José dos Campos: INPE. p. 379 - 386, Apr., 2009.

ROSENDO, J. dos S. Vegetation indices and monitoring of land use and vegetation cover in the Araguari River Basin - MG - using data from the Modis sensor. 2005. 130 p. Dissertation (Master's Degree in Geography and Land Management) - Postgraduate Program in Geography, Federal University of Uberlândia, Uberlândia. 2005.

SANTIAGO, M. M. SILVA, H. A. GALVINCIO, J. D. OLIVEIRA, T. H. Analysis of vegetation cover using vegetation indices (NDVI, SAVI and IAF) around the Botafogo-PE dam. Anais XIV Simpósio Brasileiro de Sensoriamento Remoto, Natal, Brazil, April 25-30, 2009, INPE, p. 3003-3009.

SILVA, V.; OLIVEIRA C.; RODRIGUESA, J.; ALENCAR, C.; OLIVEIRA, O. NDVI Estimation with Landsat 5-TM Satellite Image of the Cariri/CE Region. In: IX Latin American and Caribbean Congress of Agricultural Engineering - CLIA; XXXIX Brazilian Congress of Agricultural Engineering - CONBEA. July 25-29, 2010. Proceedings. Vitória, 2010.

SILVA, D. S. Use of NDVI image for temporal analysis of vegetation cover: case study: Teresópolis/RJ. *In:* SIMPÓSIO BRASILEIRO DE SENSORIAMENTO REMOTO, 14. (SBSR), 2009, Natal. Proceedings... São José dos Campos: INPE, 2009. p. 3071-3078. DVD, Online. ISBN 978-85-17-00044-7. (INPE-15962-PRE/10571). Available at: <http://marte.sid.inpe.br/col/dpi.inpe.br/sbsr%4080/2008/11.18.01.26.42/doc/3071-3078.pdf>. Accessed on: Dec. 7, 2017.

TÔSTO, S. G. et al. Geotechnologies and Geoinformation: the producer asks, Embrapa answers - Brasília, DF: Embrapa, 2014. 248 p.: il. - (Collection 500 questions, 500 answers).

VIGANÓ, H. A.; BORGES, E. F.; ROCHA, W. J. S. F. Analysis of the performance of the NDVI and SAVI vegetation indices from images. In: Brazilian Symposium on Remote Sensing. Apr.-May. 2011, Curitiba. Proceedings...Curitiba, 2011.Available at: Accessed on: 11 Apr. 2019.

WANG, J., WANG, K., ZHANG, M. Impacts of climate change and human activities on vegetation cover in hilly southern China. Ecol. Eng. 81 (2015), 451-461.

YAGCI, A.L., DI, L., DENG, M. The influence of land cover-related changes on the ndvi-based satellite agricultural drought indices. In: Geoscience and Remote Sensing Symposium (IGARSS). 2014 IEEE International, IEEE. pp. 2054-2057.

ZHENG. Y., HAN. J., HUANG. Y., FASSNACH. S. XIE. S. LV. E., CHEN. M. Vegetation response to climate conditions based on NDVI simulations using stepwise cluster analysis for the Three-River Headwaters region of China. Ecological Indicators 92 (2018) 18-29.

CHAPTER 2

QUANTIFICATION OF SOIL VEGETATION COVER MANAGED WITH TRITON USING VANT AND DIGITAL IMAGE PROCESSING

Ana Beatriz Alves de ARAÚJO;

Agricultural and Environmental Engineer, PhD, UFERSA, Mossoró/RN

Suedêmio Lima e SILVA;

Agricultural Engineer, PhD, UFERSA, Mossoró/RN.

INTRODUCTION

Agriculture has undergone a series of transformations, becoming an activity that increasingly requires management of its production processes. The growing development of new techniques linked to crop management, new equipment and more efficient inputs have led to significant gains in crop yields.

According to Far and Rezaei-Moghaddam (2018), precision agriculture is considered part of the agricultural management system, integrating information technology into crop development. Based on information technology, precision agriculture is able to analyze, manage and align profitability with sustainable concepts. This new sustainable concept of agricultural resources is used to manage temporal and spatial changes in the field. Moving faster towards sustainability will increasingly require incorporating precise practices and site-specific management into agricultural production. Understanding the possible environmental impacts caused by precision agriculture is fundamental to the application of this technology. In addition to reducing costs, it is necessary to increase yields and bring considerable environmental benefits. Increasing yields, improving economic production and offsetting costs are listed as advantages of applying precision farming technologies.

Precision agriculture for the purpose of managing agricultural inputs will provide differentiated production methods for producers and, like any other technology, allows farmers to acquire data with the aim of identifying effective variables in the potential yield of agricultural areas (FAR and REZAEI-MOGHADDAM, 2018).

An important option that has emerged for precision agriculture is the development of unmanned aerial vehicles (UAVs). Their technological development has been favoring their application in agricultural areas, mainly due to their lower cost, reduced equipment size and optimized production (JORGE; INAMASU, 2014).

Park, Lee and Chon (2018) state that advances in UAV technology in recent decades have enabled the acquisition of high-resolution, real-time aerial images for photogrammetry. UAVs are considered economical, and although their performance capacity has improved markedly with the development of technologies, photogrammetry using unmanned aerial vehicles cannot yet completely replace manned aerial vehicle photogrammetry due to technical limitations, such as: short battery life and instability related to turbulence caused by light platforms.

Dias (2012) states that analyzing the development of a crop based on the area of soil covered is an important strategy for preventing the growth of spontaneous vegetation, reducing water loss through evaporation and measuring the effect of vegetation cover on erosion control, while maintaining the productive capacity of these agricultural areas.

According to Cruz et al. (2008), various techniques can be used to analyze vegetation cover and these techniques will vary according to the study scenario, using satellite images to evaluate cover on a large scale. This process is relatively expensive and is commonly used in large areas, given the high cost of obtaining images, which makes it impossible to obtain images to carry out studies in small regions.

However, one possible approach is to use image processing techniques based on aerial photographs of the crop. This process stands out as a tool with great potential for acquiring parameters that help in decision-making, reducing observation time in the field and the influence of atmospheric conditions (JORGE; SILVA, 2009).

Among the existing programs, SisCob is a system used to analyze land cover. The images acquired are classified, making it possible to quantify changes and generate thematic maps. According to Oliveira et al. (2014), this system is based on establishing a scale of shades, defined by distinct tones and colors, which make up an artificial neural network, helping in the analysis of a selected image. Based on the recognition of the neural network (color pattern), pre-defined by the expert, the image is classified, making it possible to quantify each group formed, with the results expressed as a percentage in relation to the total area of the image.

The aim of this work is to quantify the spatial distribution of ground cover in an experimental area at UFERSA using digital images obtained by drone, using the Siscob computer program to classify the images, at three travel speeds.

MATERIAL AND METHODS

The experiment was conducted at the Rafael Fernandes Experimental Farm, located in the community of Alagoinha (5°03'37"S; 37°23'50"W and altitude 72 m), belonging to the Federal Rural University of the Semi-Arid (UFERSA). The soil in the experimental area is classified as Argissolo Vermelho-Amarelo (PEREIRA, 2011).

The experimental design was entirely randomized blocks, consisting of four treatments, unmanaged area (SM), V1 (6.0 km.h^{-1}), V2 (8.1 km.h$^{(-1)}$)) and V3 (9.7 km.h^{-1}) with nine replications, totaling 36 experimental units. The variables analyzed were: percentage of maize straw cover, percentage of spontaneous vegetation, percentage of bare soil and percentage of crop remains. The data was submitted to analysis of variance using the "F" test at 5% probability. The means were then compared using the Tukey test at 5% probability. The SISVAR 5.0 computer program (FERREIRA, 2008) was used for statistical analysis.

The area was cultivated with corn, using a conventional tillage system in a center pivot area, with a row spacing of 90 cm and a stand of 50,000 plants per hectare. Harvesting was carried out manually, removing the cobs and leaving the aerial part of the plant in place.

In order to assess the percentage of ground cover made up of maize straw and spontaneous vegetation before and after management, aerial images were taken using a DJI Phantom II drone camera with a resolution of 12 megapixels and a flight height of 30m.

To manage the maize straw, a Jan model 3600 Triton with a working width of 3.6 m was used, coupled to a Jonh Deere model 6110J tractor, operating in the unmanaged area (SM) and at three different travel speeds (6.0, 8.1 and 9.7 km.h$^{-1)}$.

The images of the experimental area were captured using a square target measuring 0.5 x 0.5 m, in order to keep the photos in the same position. The target was first positioned to capture the images and after this process, it was removed to manage the area; when the management of the area was finished, the target was placed back there to capture another image.

The photos were previously processed using the evaluation version of the graphic design software CorelDraw Graphics Suite 2018, an image editing program designed to increase the degree of definition of the images.

The images were classified using the SisCob. V.1.0 (Software for Land Cover Analysis), provided by Embrapa Instrumentação Agropecuária; with the aim of quantifying the percentage of vegetation cover per area.

To classify the images, a neural network was created with three classes: corn stover, spontaneous vegetation and bare soil. The patterns for each class were determined through selection windows in the image, guided by the different shades of color present in them.

RESULTS AND DISCUSSION

Using the computer program Siscob V.1.0, the original images were cropped and then processed, establishing three classes. After this process, the neural network was created with the predefined patterns and colors for classifying the images: corn stover = gray color, spontaneous vegetation = green color and bare soil = orange color.

Figure 1A shows a crop of the original photo with the vegetation cover before management. Figures 1B and 1C are cropped images processed with the evaluation version of the graphic design software CorelDraw Graphics Suite 2018, to increase the contrasts between the corn straw, spontaneous vegetation and bare soil in the classified image, respectively.

(a)   (b)   (c)

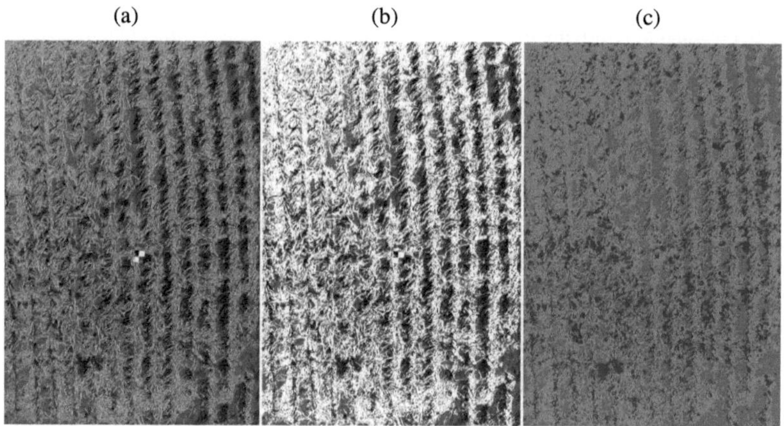

Figure 1: A) Cutout of the original photo of the area. B) Cutout of the processed image. C) Image classification using Siscob V.1.0.

In the mean test (Table 1) for the variable maize straw (MP), the treatment with no management (SM) differed significantly from the other treatments, showing that it had a lower percentage of straw. In this situation, the lack of straw (residue from the junction between maize straw and spontaneous vegetation) is due to the existence of the aerial part of the plant in the unmanaged area, which remained after the ears were harvested by hand. For the other treatments: speed of 6.0 km.h^{-1} (V1), speed of 8.1 km.h$^{(-1)}$ (V2) and speed of 9.7 km.h^{-1} (V3), the percentage of coverage did not change significantly, i.e. for all travel speeds, the soil was covered with straw.

Table 1: Completely randomized block design (DIC).

Treatments/ Variables	% Straw Maize	% Spontaneous Vegetation	% Bare Ground	% Cultural remains
SM	48,0 b	14,4 b	37,6 a	62,4 c
V1 - 6.0 Km.h^{-1})	70,8 a	19,6 a	9,5 b	91,5 a
V2 - 8.1 Km.h^{-1})	69,4 a	17.4 ab	13,2 b	86.8 ab
V3 - 9.7 Km.h^{-1})	67,4 a	15,0 b	17,7 b	82,3 b
CV %	9,6	15,1	32,7	7,9
DMS	8,0	3,3	8,3	8,3
Average	63,9	16,6	19,5	80,8

Note: SM: no management, V: speed, CV: coefficient of variation, MSD: minimum significant difference.

For the spontaneous vegetation (SV) variable, there was a significant difference between the SM treatment, which had 14% spontaneous vegetation, and the V1 treatment, which had 19% of the soil covered by particulate matter. This shows that there is more spontaneous vegetation covering the soil in the managed area. There was also a significant difference between speeds V1 and V3. In managed areas, the lower the travel speed, the better the particulate matter is distributed, improving soil coverage. According to Vitória et al. (2011) the type of preparation

soil provides differences in the percentage of plant cover incorporated. For Nagahama et al. (2016), the set with the highest plant cover was at a speed of 3.88 km h^{-1}, and the highest cover was at a speed of 2.45 km h^{-1}. This emphasizes that lower speeds have greater vegetation cover. Lopes et al. (2010) state that the highest percentage of soil cover was seen at lower travel speeds, due to the greater depth of penetration into the soil by the active organs of the "aerosol" prototype with different ballast and angles between sections.

In the bare soil variable (SD), there was no statistically significant difference between the travel speeds. The difference was between treatments (SM) and all speeds. In conclusion, the natural soil cover in the unmanaged area differed from that in the managed area. The unmanaged area has a lower percentage of soil cover than the managed area, since it distributes the crushed material evenly, favoring increased soil cover.

The differences between the SM, V1, V2 and V3 treatments occur mainly because the lower travel speeds distribute the residue more evenly, reducing the percentage of bare soil. According to Alvarenga et al. (2011), this result is considered positive, as the amount of straw on the ground and the uniformity of its distribution can serve as a reference for a preliminary assessment of the conditions in which a possible no-till system could be developed.

In the crop residue (CR) variable, the SM treatment differed from all the speeds, as it is understood that the unmanaged area has the straw on the erect corn, which remained after harvest. In the area that had soil management, this material was crushed and distributed over the soil. The V1 treatment also differed from the V3 treatment; in other words, at the lower speed, the soil was covered more evenly with the crop remains than at the higher speed. According to Alvarenga et al. (2011), these residues must cover at least 50% of the soil surface to achieve a good distribution rate. This is one of the most important requirements for the success of SPD, as it affects practically all the changes that the system promotes. According to Santos (2010), this behavior is attributed to the variation in traction force values, i.e. the lower the traction force, the higher the travel speed, which can result in a fluctuation effect. This agrees with Nagahama et al. (2013) who state that lower travel speeds favor greater working depths.

CONCLUSIONS

At all three travel speeds, the soil surface is covered with more than 50% of its area covered with straw.

The treatment with the highest straw distribution rate was V1, with 90.5% of the soil covered. Increasing travel speed increases the percentage of bare soil.

The use of drones to obtain aerial images is viable for assessing soil cover.

REFERENCES

ALVARENGA, R.; CABREZA, W.; CRUZ, J.; SANTANA, D., Soil cover plants for no-till systems. Informe Agropecuário, Belo Horizonte, v. 22, n 208, 2011.

CARMO FILHO F.; OLIVEIRA O. Mossoró: um município do semi-árido nordestino, caracterização climática e aspecto florístico. Mossoró: ESAM, (Coleção Mossoroense, Série B), 1995.

CRUZ, E., CARVALHO, D., VARELLA, C., SILVA, L., SOUZA, W., PINTO, F. Comparison of digital image classifiers in determining soil cover. Engenharia Agrícola, Jaboticabal, v. 28, n. 2, 2008.

DIAS, A. Soil cover plants in the attenuation of water erosion in the south of the State of Minas Gerais. Dissertation (master's degree) - Federal University of Lavras. Lavras, 2012.

EMBRAPA. Brazilian soil classification system. 2. Ed. Rio de Janeiro: EMBRAPA, 2006.

FAR, S.; REZAEI-MOGHADDAM, K. Impacts of the precision agricultural technologies in Iran: An analysis experts' perception & their determinants. Information Processing in Agriculture 5 (2018) 173-184.

FERREIRA, D. SISVAR: A program for statistical analysis and teaching. Revista Symposium, v.6, p.36-41, 2008.

JORGE L., SILVA, D. SisCob: user manual. São Carlos: Embrapa Instrumentação Agropecuária, 2009.

JORGE L., INAMASU Y. Use of Unmanned Aerial Vehicles (UAVs) in Precision Agriculture. Embrapa Instrumentação - São Carlos, SP, 2014.

LOPES, A., CAMARA, F.T., SCALA JÚNIOR, N.L., FURLANI, C.E.A., SILVA, R.P., BARBOSA, A. L. P. B. Operational performance of an "aerosol" prototype. Engenharia Agrícola, v. 30, n. 1, p. 82-91, 2010.

NAGAHAMA, H. J.; CORTEZ, J. W.; PIMENTA, W. A.; PATROCÍNIO, A. P.; SOUZA, E. B. Tillage systems and travel speed of mechanized units on soil attributes. Revista Agrarian, v.9, n.34, p. 357-364, Dourados, 2016.

NAGAHAMA, H. J., CORTEZ, J. W., PIMENTA, W. A., PATROCÍNIO FILHO, A.P., SOUZA, E. B. PERFORMANCE of the tractor-equipment combination in a periodic tillage system on Yellow Argissolo. Revista Energia na Agricultura, v. 28, n. 2, p. 79-89, 2013.

OLIVEIRA L, *MATSUMOTO, S., SILVA, R., SILVA, V., OLIVEIRA, P.* Methods for quantifying and interpreting the spatial distribution of ground cover in wooded coffee plantations. Coffee Science, Lavras, v. 9, n. 2. 2014

PARK, S., LEE, H., CHON, J., Sustainable monitoring coverage of unmanned aerial vehicle photogrammetry according to wing type and image resolution. Environmental Pollution (2018), doi: 10.1016/j.envpol.2018.08.050.

PEREIRA, V., SOBRINHO, J., OLIVEIRA, A., MELO, T., VIEIRA, R. Influence of El Nino and La Nina events on rainfall in Mossoró-RN. Encyclopedia Biosfera. Centro Científico Conhecer, v. 7, 2011.

SANTOS, M.S. Mechanical traction parameters in watermelon cultivation systems. Santa Maria: Federal University of Santa Maria, 2010. 97f. Master's dissertation in Agricultural Engineering.

VITORIA, E.L.; LONGUI, F.C.; FERNANDES, H.C.; GUIMARÃES FILHO, C.C. Influence of the type of soil preparation and sowing speed on agronomic characteristics of the corn crop. Revista Agrotecnologia, v. 2, n. 2, p. 44-52, 2011.

CHAPTER 3

ECOMPOSITION OF PLANT ORGANIC MATTER IN SEMIARID SOILS

Erika Socorro Alves Graciano de VASCONCELOS;

PCI researcher at the National Semi-Arid Institute - INSA. Plant production area.

Evaldo dos Santos FELIX;

PCI researcher at the National Semi-Arid Institute - INSA. Plant production area.

Daniela Batista da COSTA;

PCI researcher at the National Semi-Arid Institute - INSA. Plant production area.

Kaline Dantas TRAVASSOS;

PCI researcher at the National Semi-Arid Institute - INSA. Water Resources Area

Jucilene Silva ARAÚJO;

Full Researcher at the National Semi-Arid Institute - INSA. Plant production area.

INTRODUCTION

The decomposition of organic matter (OM) of plant origin, also known as plant residue, in the soils of the semi-arid tropics is a complex process that acts to maintain the functionality of the ecosystem, allowing part of the carbon incorporated into the organic matter to return to the atmosphere in the form of CO_2 and another part, together with the chemical elements present in its constitution, to be incorporated into the soil, becoming part of the soil organic matter (SOM) and mineral particles, as a component of the soil.

Essentially, the decomposition of plant residues involves the transformation of the state of a given residue, under the influence of biotic and abiotic regulatory factors inherent in the environment. The effectiveness of this process depends on the interaction of three factors: the characteristics of the plant residue, the decomposer community (biotic factors) and the edaphoclimatic conditions of the environment (abiotic factors).

The description of these factors plays an important role in the management of cultivated areas in the semi-arid region, which can help to develop techniques that improve the use of plant residues in the supply of organic matter and nutrients to the soil. This chapter presents a bibliographical review of the main factors influencing the decomposition process of plant organic matter in semi-arid soils.

Characteristics of plant residues that influence decomposition

The characteristics of plant residues, such as their qualitative composition; the chemical structure of the constituents, in terms of the basic unit, types of chemical bonds, size, shape and degree of polymerization; their functions in plant tissue, structural and reserve; age and particle size, are important factors in the decomposition of plant residues as they have a direct influence on the speed of decomposition (Moreira & Siqueira, 2006; Araújo et al., 2008), thus determining the degradability of the residue (Gama-Rodrigues et al., 2007). These characteristics can change between residues from different plant species and within the same species.

The quality of plant residues has been defined by their chemical characteristics, shown in Table 1, such as carbon, nitrogen, phosphorus and sulphur content; the types and proportions of compounds present in the residue such as cellulose, pectin, hemicellulose, lignin, polyphenols, proteins, amino acids, starch, sugars, among others, and by some ratios between some compounds such as Carbon:Nitrogen (C:N), Carbon:Phosphorus (C:P), Lignin:Nitrogen (Lignin:N), Polyphenols:Nitrogen (POl:N) and Lignin + Polyphenols:Nitrogen (Lig+Pol:N).

Table 1. Chemical characteristics of vegetable waste.

Species	N	P	K	C:N	Polyphenol	Lignin	Hemicellulose	Cellulose
	------ g.kg⁻¹ -			----------------------------- % ---------------------				
Leucaena	40,17	1,55	11,18	12	18,0	9,72	12,93	14,52
Guandu	28,75	2,83	8,22	18	9,0	16,19	18,0	15,77
Shadowman	22,71	1,30	6,75	23	17,6	17,47	15,97	21,66
Acacia	18,28	0,51	5,4	27	21,4	14,28	16,61	17,43
Palma	11,4	1,4	27,8	—	—	4,62	3,30	21,15

Source: Loss et al., 2009; Tosto et al., 2007; Teles et al., 2004.

The content of N, P and S in plant residues is extremely variable, depending on the species of residue and, among other things, soil fertility, which can limit the decomposition of plant residues. On the other hand, the total C content is not very variable, as it forms part of the skeleton of the basic biomolecules (proteins, polysaccharides, nucleic acids and lipids) that make up organic compounds (Redin, 2010).

Each of the compounds present in plant residue, mentioned above, has specific characteristics related to its composition, which are fundamental in the decomposition process, so there is a need to know them. To this end, the specific characteristics of some compounds and their decomposition in the soil are described below:

Cellulose is a resistant, water-insoluble polysaccharide made up of chains of glucose units connected by β-1,4-glycosidic bonds, which give it characteristic structural properties (Figure 1), joined together by hydrogen bonds (Ogeda & Petri, 2010), with a highly ordered crystalline structure and high molecular mass (Carvalho et al., 2005). In cellulose, the glucose units form straight, extended chains which are arranged side by side, producing a fibre structure stabilized by intra- and inter-chain hydrogen bonds, as shown in figure 2. This fibre structure gives cellulose greater strength. This compound accounts for most of the CO_2 fixed by plants, making it the main component of plants. In the soil, this molecule is broken down by cellulase enzymes produced by cellulolytic microorganisms, which break down its high molecular weight molecule into cellobiose (a disaccharide made up of glucose linked to glucose) and free glucose. Under aerobic conditions, glucose is oxidized via the tricaboxylic acid cycle, while anaerobic fermenters produce acetate, propionate, butyrate, H_2 and CO_2 from glucose (Moreira & Siqueira, 2006).

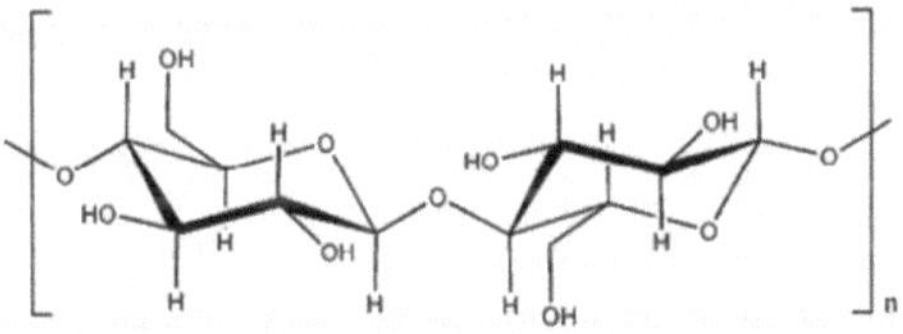

Figure 1: Molecular structure of cellulose (Pinto, 2008).

Pectins form a complex group of structural polysaccharides found in the primary cell wall and intercellular layers of terrestrial plants, which are associated with cellulose, hemicellulose and lignin and are abundant in young tissues and fruits. This polysaccharide contributes to adhesion between cells and to the mechanical strength of the cell wall. Structurally, its molecules (figure 3) are made up of a linear main chain of repeated units of

(1→4)-α-D-galacturonic acid, part of which is esterified as a methyl ester (Brandão & Andrade, 1999). Its decomposition in the soil (figure 4) occurs by pectidase, represented by the enzymes protopectinase, pectin methylesterase and polygalacturonase, which act in the degradation of pectic substances, which are produced by bacteria of the genus Erwinia, Clostridium, Peseudomonas and Bacillus (Moreira & Siqueira, 2006).

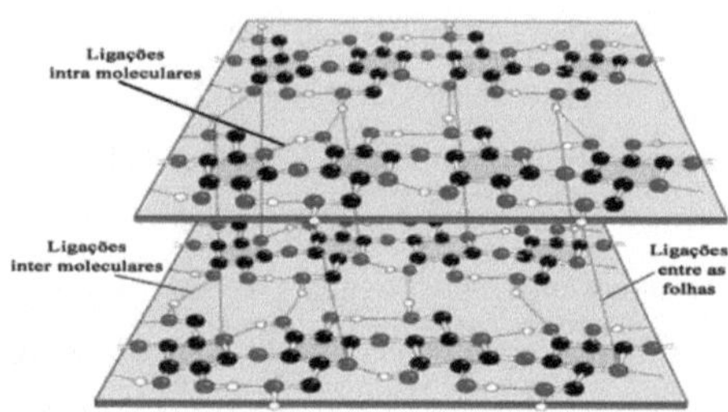

Figure 2 - Cellulose structure showing the hydrogen bridge bonds between the sheets (blue), intra molecular (red) and inter molecular (green) (Pinto, 2008).

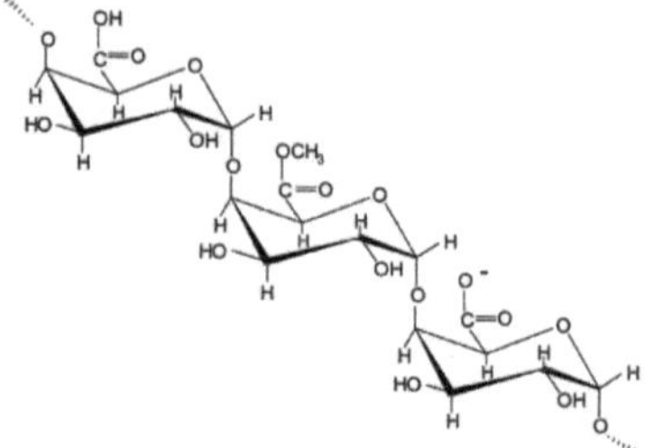

Figure 3. Chemical structure of the pectin chair (Source: Brandão & Andrade, 1999).

Hemicelluloses consist of branched polymers made up of polysaccharides (figure 5) with a low degree of polymerization (Ferreira & Rocha, 2009) and low molecular mass, associated with the plant cell wall (Carvalho et al., 2005), 2005), it is the second largest component of plants, made up of pentose arrangements, such as xylose and arabinose; hexose, such as mannose, glucose and galactose; and sometimes uronic acids, such as glucuronic and galacturonic (Silva et al., 2009). Hemicelluloses are not crystalline and are interconnected to the cellulose microfibrils, providing elasticity and preventing them from touching (Ferreira & Rocha, 2009). Its decomposition can be hindered when it is linked to another substance, such as hemicellulose fibrils, when they form hydrogen bridges with fibers in the plant cell wall matrix (Moreira & Siqueira, 2006).

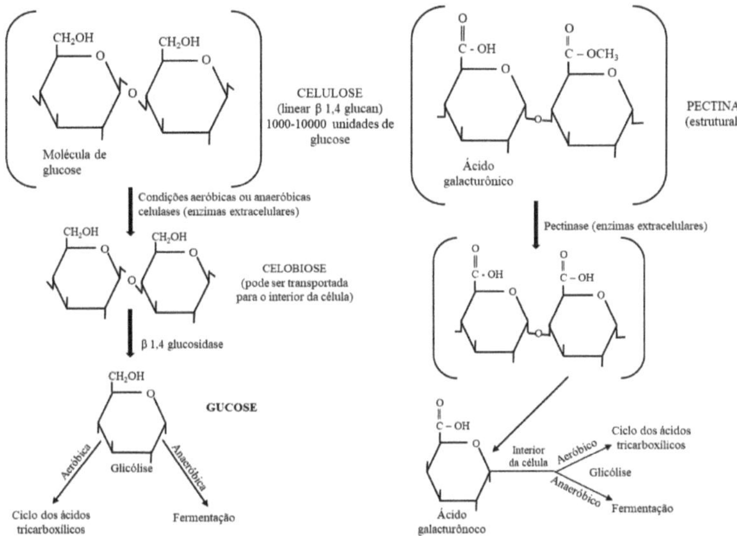

Figure 4: Decomposition of cellulose and pectin in the soil (Source: Moreira and Siqueira, 2006).

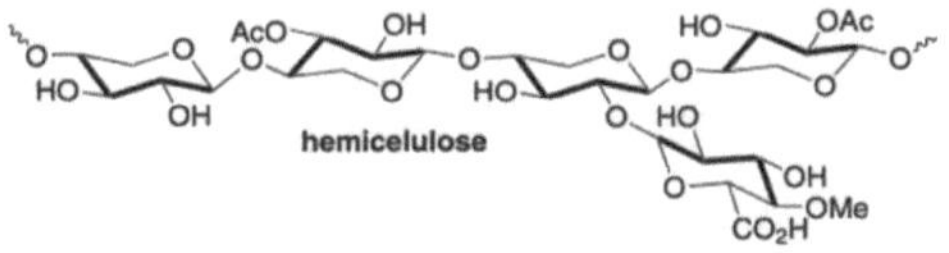

Figure 5 - Partial structure of hemicellulose (Ferreira & Rocha, 2009).

Lignin is a hydrophobic macromolecule with a three-dimensional, highly branched structure (figure 6) and can be classified as a polyphenol, made up of an irregular arrangement of various phenylpropane units which may contain hydroxyl and methoxyl groups as substituents on the phenyl group (Silva et al., 2009), which is responsible for 25% of the dry phytomass produced annually in the biosphere (Moreira & Siqueira, 2006). It is composed of a set of high molecular mass amorphous reticular polymers, most often associated with cellulose and hemicellulose, with a strongly aromatic chemical structure, combined with benzene rings containing free and methylated phenolic groups (Carvalho et al., 2005), which contributes to its high recalcitrance. Ether bonds dominate the bonding between the lignin units. The strength of adhesion between cellulose fibers and lignin is increased by the existence of covalent bonds between the lignin chains and the constituents of cellulose and hemicellulose (Silva et al., 2009). It is located in the cell walls of various types of supporting and vascular tissues (Taiz & Zeiger, 2009). The concentration of this constituent in plant residues has been considered to be one of the most important factors in the speed of decomposition in terrestrial ecosystems, which has made parameters that include the content of this constituent good benchmarks of decomposition dynamics (Correia & Andrade, 2008).

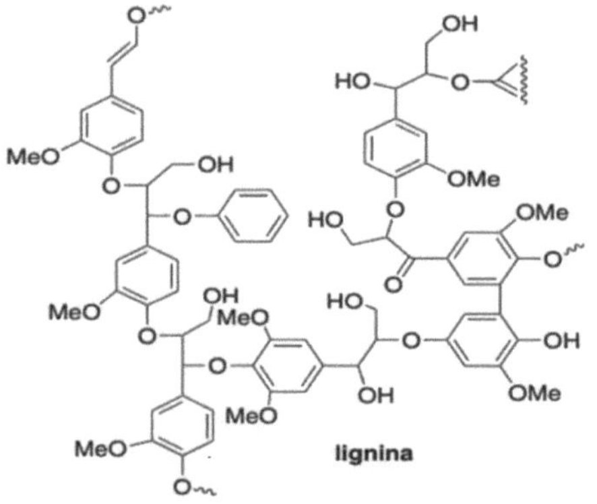

Figure 6 - Partial structure of lignin (Ferreira & Rocha, 2009).

In the soil (figure 7), lignin is degraded by specific groups of Basidiomycetes belonging to the order Agaricales and by some Ascomycetes. By attacking polysaccharides associated with lignin, these specific groups remove the CH_3 and $R-O-CH_3$ side-chains from the lignin, leaving phenols which, when oxidized, become brown or brownish, which condense and can form humic substances (Moreira & Siqueira, 2006).

Polyphenols are secondary aromatic compounds derived from phenols that make up around 5 to 15% of plant weight and are often analyzed in plant leaves. Many of these secondary compounds are released as tannins, which cause protein precipitation. Tannins, such as gallic and protocatenoic acids, impair the decomposition process and can influence the speed of decomposition of plant residues (Auer et al., 2006).

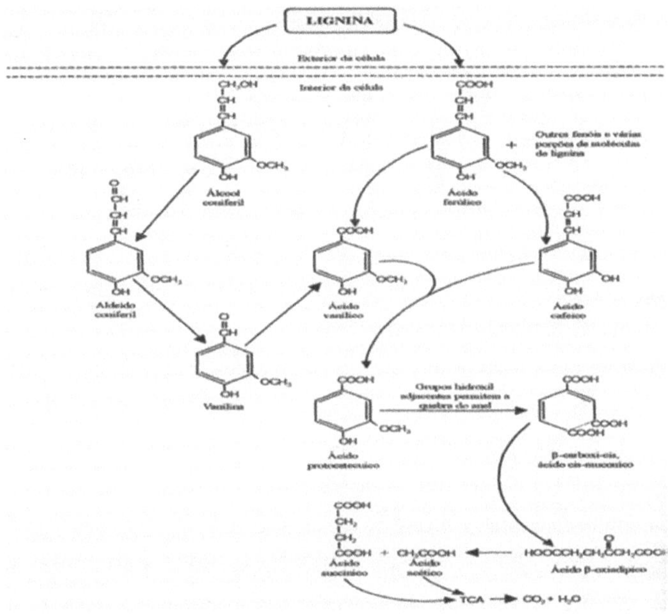

Figure 7. Lignin degradation route in soil (Paul & Clark, 1996).

Starch is a compound made up of a mixture of two glucose polymers: amylose and amylopectin, which represents the most important reserve organic compound in plants. The differences between these polymers lie in the chaining of the glycosidic units, where in amylose the chain is linear and the bonds are α-($1\rightarrow4'$) and in amylopectin there is a branch in the C6 hydroxyl with an α-($1\rightarrow6'$) bond (Ferreira & Rocha, 2009). Few studies have been carried out on its decomposition in the soil. Among the degraders of this compound are actinomycetes, which produce organic acids, CO_2 and dextrins during decomposition (Moreira & Siqueira, 2006).

Lipids are complex esters of fatty acids and alcohols, and little is known about their degradation, but it is known that bacteria mainly attack natural fats and waxes (Moreira & Siqueira, 2006).

Proteins, on the other hand, are compounds with a higher nitrogen content that are easily broken down by various microorganisms that produce enzymes in plant residues and soil. However, proteins can be associated with other compounds such as lignins, tannins and even clays. This association increases the resistance of proteins to decomposition and, as a result, they can remain in the soil (Moreira & Siqueira, 2006).

The degradation of this set of organic compounds occurs through specific metabolic routes, producing different intermediate substances that interact differently with other compounds and, above all, with the macro and microorganisms that decompose them, so that each compound is broken down differently in the soil. Therefore, through the composition and specific characteristics of the compounds, it can be seen that plant residues are made up of compounds that are difficult to degrade, i.e. more resistant to biochemical degradation, such as lignin, and those that are easier to degrade, such as carbohydrates, proteins, hemicellulose and cellulose, which break down more quickly, constituting the substances of immediate importance for decomposition.

The degradation of plant residue can occur more slowly or more quickly depending on the higher or lower concentration of these compounds in the residue. An example of this is the decomposition of younger plant residues, which occur more quickly, while older residues decompose more slowly, mainly due to the chemical composition of the residue. Young residues generally contain higher levels of sugars, starch, amino acids and proteins, which have a low molecular weight and are easy to decompose, while older residues have higher levels of hemicellulose, polyphenols and lignin and decompose more slowly (Leite & Gavão, 2008). Therefore, compounds that decompose slowly or are not decomposed at all are called recalcitrant (Moreira & Siqueira, 2006)

The decomposition rate of plant residues is also associated with the ratios of carbon:nitrogen (C:N), carbon:phosphorus (C:P) and carbon:sulphur (C:S) in the residues. These ratios are commonly used as indicators of plant residue decomposition (Souto et al., 2009; Alves et al., 2011), as they markedly influence the availability of N, P and S in the soil according to the values of the ratios. Depending on the C:N ratio of the plant residue, if the C:N ratio is high (>30) the N may be depleted, and if the ratio is low (<20) the mineralized element may be released. These ratios are higher for P and S (Table 2), due to the microorganisms' requirements for a smaller amount of these nutrients in relation to C (Moreira & Siqueira, 2006).

Table 2. Relationships between the initial quality of the substrate and the unit ratios of C, N, P and S that determine the availability of nutrients in the soil.

Waste quality	Relationship			Immobilization (I)/ Mineralization (M)	Nutrient availability
	C:N	C:P	C:S		
Poor	>30	>300	>400	I > M	Decreased
Intermediate	20-30	200-300	200-400	I = M	Unchanged
Rico	<20	<200	<200	I < M	Increased

Source: Moreira & Siqueira, 2006.

Plant residues from agricultural crops can have C:N ratios of between 10 and 100, as is the case with corn straw, which has a C:N ratio of 69:1 (Silva et al., 2008), sugarcane straw of 34:0.8 (Vitti et al., 2008) and leguminous plant residues of around 20:1 or less, which influences the rate of decomposition. Although the C:N ratio has been widely used as a quality variable for decades to predict decomposition and mineralization processes, it is not always favourable as it does not reveal how C and N are distributed among the different chemical compounds that make up plant residues (Alves et al., 2011; Oliveira et al., 2012).

In addition to the C:N, C:P and C:S ratios, the lignin:N, polyphenols:N and lignin+polyphenols:N ratios of plant residue are important and are also used as indicators of decomposition (Rheinheimer et al., 2000; Aita & Giacomini, 2003; Espindola et al., 2006; Silva et al., 2009). As these ratios increase, decomposition of the residue decreases and the availability of N in the short term may also decrease.

According to Alves et al. (2011), studying the relationship between quality and N release by semi-arid plants used as green manure, they observed that materials with polyphenol:N ratios below 0.5 cause net mineralization of N and ratios above 5 cause immobilization. The same authors report that the interference of phenolic compounds on the N decomposition process occurs because polyphenols form complex structures through stable bonds (hydrogen bridges, covalent bonds, among others) with nitrogen groups, making the materials more resistant to decomposition.

The quality of plant residues in the soil is a factor of great importance as it directly influences the speed of the decomposition process and has therefore been the focus of a number of studies. Zhang et al. (2008), evaluating the decomposition of plant residues in terrestrial systems, indicated that the decomposition rate of the residue increases with an increase in N, P and K content, and decreases with an increase in lignin content and the C:N and lignin:N ratios. Souto et al. (2009), analyzing the chemical characteristics of plant residues deposited in caatinga areas, observed that high C:N and C:P ratios of plant residues showed a lower decomposition rate. Berg & McClaugherty (2003), studying the decomposition of plant residues, reported that the concentrations of N and P regulate decomposition rates in the initial phase of the decomposition process of plant material, while the concentration of lignin has a greater influence in the more advanced stages of decomposition. Residues with high levels of lignin, cellulose and polyphenols and high C:N and C:P ratios correspond to materials with lower decomposition rates and a high capacity to remain in the soil (Leite et al., 2010). The influence of plant residue characteristics on the speed of decomposition is related to biotic factors in the environment, especially the stimulus it exerts on the activity of decomposer organisms.

Biotic factors that influence the decomposition of organic matter

The decomposition of plant residues is essentially a biological process (Correia & Andrade, 2008), with the organisms that make up the fauna and microorganisms that inhabit the soil being the main biotic agents of the physical and chemical disruption that occurs in the decomposition of plant residues added to the soil, characterizing them as an essential factor in this process.

The components of the soil fauna are classified according to their size (Table 3) into microfauna, mesofauna and macrofauna. The organisms that make up the microfauna vary in body diameter from 4 µm to 100 µm and include protozoa, rotifers, copepods, tardigrades, nematodes and others. The mesofauna includes organisms with a body diameter of between 100 µm and 2 mm, including mites, springtails, some groups of myriapods, arachnids and various orders of insects, some oligochaetes and crustaceans. In turn, the organisms of the soil macrofauna have a body diameter of between 2 and 20 mm and can belong to almost all the orders found in the mesofauna, except mites, springtails, protists and diplurids (Correia and Andrade, 2008).

Soil microorganisms, also known as microbiota, represent the greatest biological and physiological diversity in the soil, as well as the greatest decomposition power (over 95%) (Leite and Galvão, 2008). The main microorganisms are bacteria, actinomycetes, fungi and algae. These organisms make up 1 to 4% of the total carbon and occupy around 5% of the soil's pore space alone, despite having a high diversity and quantity. Their presence in a given soil is a function of the prevailing environmental conditions such as aeration, humidity, pH, nutrient availability (C, N, P and S) and soil texture (Roscoe et al., 2006), which means that few microorganisms are active in the soil.

The fraction of living and most active microorganisms in the soil is also called microbial biomass, with bacteria and fungi being the most relevant, accounting for around 90% of soil microbial activity (Andreola & Fernandes, 2007). These microorganisms use enzymatic apparatus responsible for various synthesis and degradation mechanisms in the soil, sometimes promoting the mineralization of organic compounds and the release of nutrients, sometimes immobilizing them in their biomass (Leite & Galvão, 2008).

The activity of fragmenting plant residues is one of the most important functions carried out by soil fauna. The fragmentation carried out by the crushing action of the fauna reduces the size of the plant residue and increases the contact surface, which favors the action of microorganisms in the process of decomposition/mineralization of the residue deposited on the soil and in the transformation of the soil's organic matter and, consequently, in the formation of humic substances (Silva & Mendonça, 2007).

The fragmentation carried out by the ingestion of the waste by soil fauna organisms, although there are no major chemical changes, also promotes major physical changes, providing a reduction in size with an increase in the relative contact surface and hydration. These characteristics, combined with the substantial increase in the amount of microorganisms that occurs during the passage through the digestive tract of the soil fauna, make the waste particles points of great action for microorganisms, promoting decomposition even after a few days of deposition. This procedure is of considerable importance in the decomposition of plant residues, especially those of low quality (Correia & Andrade, 2008).

The activity of transforming plant residues is the main function carried out by soil microorganisms, which involves four distinct groups of transforming microorganisms: cellulolytic, hemicellulolytic, pectinolytic and ligninolytic (Tauk, 1990). The set of transformations and chemical reactions catalyzed biologically and metabolically by them is called microbial activity. This activity can be evaluated as an index of total soil activity by respiration, which is one of the oldest parameters that quantifies its activity, represented by the oxidation of organic matter by aerobic soil organisms, using O_2 as an electron acceptor, to $CO_{(2)}$, and can be analyzed by O_2 consumption or $CO_{(2)}$ production (Moreira & Siqueira, 2008).

Table 3. Size*, density and biomass of the main groups of soil organisms per m^2.

Groups	Biomass	Density
	------- g ------	-------Número***------
Microorganisms**		
Bacteria	50	10^{14}
Fungi	100	10^{11}
Algae	1	10^{10}
Microfauna (4 µm to 100 µm)		
Protozoa	9	10^{10}
Rotifers	0,01	10^5
Nematodes	10	10^7
Mesofauna (100 µm and 2 mm)		
Mites	1	10^5
Pigeons	0,6	10^4
Enquitreides	2	10^4
Macrofauna (2mm and 20mm)		
Worms	50	10^2
Termites	4,0	10^3
Ants	0,5	10^3
Coleoptera	2,5	10^2
Arachnida	0,5	10
Myriapoda	1,5	10
Others	2,5	10^2

* Based on body width. ** Includes actinomycetes. *** Order of magnitude. Source: Leite & Galvão, 2008.

In general, the components of the fauna and the microorganisms act in an interactive way, forming an intense trophic chain in which the regulators are responsible for crushing plant waste and the microorganisms for transforming it (Moreira and Siqueira, 2006). The waste deposited is initially broken down by fauna organisms and later by microorganisms, where part of the carbon present in the waste is released into the atmosphere in the form of CO_2, and the other part becomes organic matter (Leite & Galvão, 2008).

The degradation of a complex waste is processed more efficiently in an environment where there is a community of fauna and microorganisms than in the presence of a single population, because there are different species capable of degrading the different compounds that make up the waste. The occurrence and activity of a community of decomposer organisms in a given soil is a function not only of the characteristics of the plant residues (Botelho et al., 2001), but also of the region's soil and climate factors. These factors can limit or favor the survival and activity of decomposer organisms and are thus interconnected with the process of decomposition of plant residues.

Abiotic factors that influence the decomposition of organic matter

The main environmental factors, related to the physical and chemical properties of the soil, which interfere in the decomposition process of plant residues in the tropical region are: texture, aeration, humidity and pH (Mantovani et al., 2006, Messias & Silva, 2008; Leite & Galvão, 2008), and the climatic factors are temperature and precipitation (Moreira & Siqueira, 2006; Sanches et al., 2009).

Soil texture, defined as the relative proportion of granulometric fractions of less than 2 mm in size, corresponding to the sand, silt and clay fractions that make up the soil mass, is one of the most important properties in helping to explain agricultural land management, through its association with other soil properties (Nascimento et al., 2003). Texture influences the soil's physical attributes by modifying the size distribution and continuity of pores and the stability of aggregates, factors which alter the availability of water in the soil, the diffusion of gases, the movement of soil organisms, microbial access to physically protected organic matter (Leite & Mendonça, 2003) and, consequently, the presence of microorganisms which decompose plant residues in the soil.

The influence of texture on decomposition depends mainly on the amount and type of clay present in the soil, which may or may not favor microbial growth, especially during the initial phase of degradation of available waste. Soils with a clayey texture, i.e. with a higher proportion of clay, are able to retain more MO and water, however, the diffusion of gases may be lower, which are required by microorganisms in the oxidation reactions of decomposition. On the other hand, soils with a predominance of 2:1 type clay, made up of two sheets of Si tetrahedrons joined by one of Al octahedrons, have a greater capacity to retain nutrients and water at a higher adsorption force than the 1:1 type, because this type of clay has more charge, providing greater stability to the organic fraction for decomposition by microorganisms, which is a limiting factor for the development of decomposing microorganisms.

The protection that clays provide to OM is explained by the high specific surface area that determines the great interaction between clay minerals and OM, resulting in greater stability of the organic fraction to the decomposition process by microorganisms (Araújo et al., 2008). Thus, as the proportion of clay increases, the surface area of the soil's mineral matrix and the potential for stabilizing organic matter increase, affecting the action of decomposing microorganisms.

Soil aeration, which refers to the diffusion of atmospheric gases in the soil profile, mainly ensures oxygenation of the soil profile, which is necessary for aerobic microbial life. In soil, the oxygen content can be lower than that found in the atmosphere (21%), and most biological activities are impaired if the oxygen content is below 15% (Lier, 2010). This can be seen in the oxidation reactions carried out by microorganisms that decompose plant waste, which are dependent on the presence of oxygen (O_2). In this sense, any interference with soil aeration can influence the activity of microorganisms, thus affecting decomposition. A clear example of this interference is soil disturbance caused by plowing and harrowing, which accelerates the decomposition of plant residues. The absence of soil disturbance, observed in a no-till farming system, reduces decomposition by keeping plant residues on the soil surface for longer.

Soil moisture, in terms of the amount of water present in the soil, is indirectly associated with aeration, where these fractions (water and air) complement each other and together fill the soil's pore space, and also has a direct influence on the activity of the plant residue decomposer community, as many microfauna organisms live associated with the soil's water film (Correia & Andrade, 2008) and all living organisms need water to carry out their activities. Thus, the activity of decomposers can be slowed down by extreme humidity conditions, whether reduced or saturated. During periods of low water availability in the soil (water deficiency), osmotic fluctuations result in a reduction in microbial populations or a reduction in their growth. In the case of soils with high water availability (waterlogged soils) and insufficient drainage, microbial activity is altered so that the decomposition of plant residues becomes slower, and there may be an accumulation of plant residues on the soil surface and even organic matter in the surface horizons of the soil profile.

The hydrogen potential (pH) of the soil indicates a condition of acidity, neutrality or alkalinity (pH < 7; pH = 7 and pH > 7). These conditions can favor or hinder the development, activity and distribution of some microorganisms by influencing most of the reactions in the soil, which is a determining factor in the process of decomposing plant residues. Soil pH can vary greatly in a given environment, which influences the survival of different microorganisms. Under the different soil pH conditions, you can find microorganisms that thrive best in acidic conditions (acidophiles), those that thrive in alkaline conditions (basophiles), those that cannot tolerate acidity or alkalinity (neutrophiles) and those that tolerate a wide pH range (insensitive). In general, fungi are better adapted to pH values lower than 5.0 (acidophils) and bacteria to pH values between 6.0 and 8.0 (neutrophils and basophils) (Araújo, 2008).

Most tropical and subtropical soils have a low pH and are therefore characterized as acidic (pH < 7). This is mainly due to the leaching of basic cations (Ca+2, Mg+2, k+ and Na+) and the increase in hydrogen (H+) in regions where rainfall is high. This acidity affects many oxidation reactions in the decomposition process of plant residues, which require high levels of exchangeable bases. Consequently, low (acidic) soil pH causes a decrease in microbial activity and a reduction in the decomposition of soil organic matter.

Climatic factors also have an influence on the process of decomposition of plant residues in the soil, with temperature and humidity being the decisive variables in this process, since these variables interact at the same time and influence the action of decomposer organisms. Temperature is a factor that can become decisive for the development of soil microorganisms and is related to biochemical processes in the soil. Its primary source is solar radiation, which is transformed into heat. Temperature is a variable that shows the most consistent difference between regions. In a given environment, temperature can fluctuate depending on latitude, altitude and season. On average, in tropical regions, the temperature is 15°C higher than in temperate regions and, in addition, they do not have winters with such low temperatures (Leite & Galvão, 2008).

Soil microorganisms thrive in a wide temperature range, but within this range there is one considered optimal that favors their development and activity in the soil in the decomposition process, which is around 30 to 35°C (Moreira & Siqueira, 2006), which favors greater decomposition. The response of decomposition to temperature can be measured by the Q10 coefficient, which generally occurs in the order of 2.0 in the 5 to 35°C range, i.e. the speed of decomposition in this range doubles when the temperature is raised by 10°C (Moreira & Siqueira, 2006).

High temperatures increase the speed of biochemical processes (Leite & Galvão, 2008). Thus, the rates of microbial activity in the degradation of plant residue increase with increasing temperature (Reis & Rodella, 2002) and suffer a sharp decrease at temperatures below 25°C and above 35°C. Temperature can therefore be considered an important factor that is closely related to the decomposition process, which should be taken into account, especially in tropical regions.

Water is considered a vital factor for all living organisms, so the region's rainfall regime is also a determining factor in decomposition reactions. In some tropical ecosystems, such as central Amazonia, there is evidence that decomposition rates are affected by seasonal variations, forming distinct patterns in the rainy and dry seasons (Sanches et al., 2009). Annual rainfall of 1,000 to 3,000 mm combined with high temperatures in the humid tropics results in greater decomposition. In these regions, the high rates of decomposition are counterbalanced by the higher production of plant residues that are deposited on the soil.

However, climatic factors and the physical and chemical conditions of the soil can maximize the microbiological activity of decomposers, especially when the conditions are between 30 and 35°C, soil moisture close to field capacity and adequate soil aeration (Moreira and Siqueira, 2006).

FINAL CONSIDERATIONS

The decomposition of plant residues in the soil is a process in which many factors are closely related, with each factor having a defined and fundamentally important function, which makes its magnitude complex. Among them, soil decomposer organisms can be considered the most admirable in the decomposition process, as they act throughout the transformation of the residue and respond immediately to the other factors involved.

REFERENCES

AITA, C.; GIACOMINI, S.J. Decomposition and nitrogen release from crop residues of single and intercropped ground cover plants. Rev. Bras. Ci. Solo, 27, p.601-612, 2003.

ALVES, R.N.; MENEZES, R.S.C.; SALCEDO, I.H.; PEREIRA, W.E. Relationship between quality and N release by semi-arid plants used as green manure. Revista Brasileira de Engenharia Agrícola e Ambiental, 5, p.1107-1114, 2011.

ANDREOLA, F.; FERNANDES, S.A.P. Soil microbiota in organic agriculture and crop management. In: SILVEIRA, A.P.D.; FREITAS, S.S. Microbiota do solo e qualidade ambiental, p.21-37, 2007.

ARAÚJO, A.S.F. Soil microbial ecology. In: ARAÚJO, A.S.F.; LEITE, L.F.C.; NUMES, L.A.P.L.; CARNEIRO, R.F.V. Matéria orgânica e organismos do solo, Teresina:EDUFPI, 2008, p. 47-64.

ARAÚJO, A.S.F.; LEITE, L.F.C.; NUMES, L.A.P.L.; CARNEIRO, R.F.V. Matéria orgânica e organismos do solo, Teresina:EDUFPI, 2008, 220p.

AUER, C.G.; GHIZELINI, A.M.; PIMENTEL, I.C.; BIZI, R.M. Fungi in leaf litter of Pinus taeda L. in stands of different ages. Floresta, Curitiba, PR, v.36, p.433-438, 2006.

BRAMDÃO, E.M.; ANDRADE, C.T. Influence of structural factors on the gelling process of highly methoxylated pectins. Polymers: Science and Technology, p.38-44, 1999.

BOTELHO, S.A.; LEANDRO, W.N.; COSTA, J.L.S. Suppressiveness induced to Rhizoctonia solani Kiihn by the addition of different plant residues to the soil. Pesq. Agrop. Tropical, v.31, p.35-42, 2001.

CARVALHO, G.B.M.; GINORIS, Y.P.; CÂNDIDO, E.J.; CANILHA, L.; CARVALHO, W.; SILVA, J.B.A. Study of eucalyptus hydrolysate in different concentrations using vacuum evaporation for fermentation purposes. Rev. Analytica, p.54-57, 2005.

CORREIA, M.E.F.; ANDRADE, A.G.; Formation of litter and nutrient cycling. In: SANTOS, G.A.; SILVA, L.S.; CANELLAS, L.P.; CAMARGO, F.A.O. Fundamentals of soil organic matter: tropical and subtropical ecosystems. 2ed. Porto Alegre: Metropole, p.561-569, 2008.

ESPINDOLA, J.A.A.; GUERRA, J.G.M.; ALMEIDA, D.L.; TEIXEIRA, M.G.; URQUIAGA, S. Decomposition and release of accumulated nutrients in perennial herbaceous legumes intercropped with banana. Rev. Bras. Ciência do Solo, 30, p.321-328, 2006.

FERREIRA, V.F.; ROCHA, D.R. Potentialities and opportunities in the chemistry of sucrose and other sugars. Química Nova, v.32, p.623-638, 2009.

GAMA-RODRIGUES, A.C.; GAMA-RODRIGUES, E.F.; BRITO, E.C. Decomposition and release of nutrients from cover crop residues in red-yellow Argissolo in the northwestern region of Rio de Janeiro. Rev. Bras. Ci. Solo, v.31, p.1421-1428, 2007.

LEITE, L.F.C.; GALVÃO, S.R.S. Soil organic matter: functions, interactions and management. In: ARAÚJO, A.S.F.; LEITE, L.F.C.; NUMES, L.A.P.L.; CARNEIRO, R.F.V. Matéria orgânica e organismos do solo, Teresina:EDUFPI, p. 11-46, 2008.

LEITE, L.F.C.; FREITAS, R.C.A.; SAGRILO, E.; GALVÃO, S.R.S. Decomposition and nutrient release of plant residues deposited on Yellow Latosol in the Maranhão savanna. Rev. Ciências Agronômica, v.41, p.29-35, 2010.

LEITE, L.F.C.;MENDONÇA, E.S. Century model of soil organic matter dynamics: Equations and pressuports. Ciência Rural, v.33, p.679-686, 2003.

LIER, Q.J.V. Soil physics. SBCS, Viçosa, 2010, 298p.

LOSS, A.; PEREIRA, M.G.; FERREIRA, E.P.; SANTOS, L.L.; BEUTLER, S.J.; FERRAZ JÚNIOR, A.S.L. Oxidizable fractions of organic carbon in red-yellow clay loam under alley system. Revista Brasileira de Ciencia do Solo, 33:867-874, 2009

MESSIAS, A.S.; SILVA, C.A.A. Microorganisms that degrade solid waste. In: Figueiredo, M.V.B.; Burity, H.A.; Stamford, N.P.; Santos, C.E.R.S. Microorganisms and Agrobiodiversity: the new challenge for agriculture. Guaíba: Agrolivros, p.423-437, 2008.

MONTOVANI, J.R.; FERREIRA, M.E.; CRUZ, M.C.P.; BARBOSA, J.C.; FREIRIA, A.C. Mineralization of carbon and nitrogen from urban waste compost in Argissolo. Rev. Bras. Ci. Solo, v.30, p.677-684, 2006.

MOREIRA, F.M.S.; SIQUEIRA, J. O. Microbiologia e Bioquímica do Solo. 2 ed. Lavras: editora UFLA, 729p, 2006.

NASCIMENTO, G.B.; PEREIRA, M.G.; ANJÕS, L.H.C.; SOARES, E.D.R.; SOUZA, M.R.P.F. Determining the textural class of soil samples using an electronic spreadsheet. Rev. Univ. Rural, Sér. Ci. Vida, v.23, p.27-30, 2003.

OGEDA, T.L.; PETRI, D.F.S. Enzymatic hydrolysis of biomass. Química Nova, v.33, p.1549-1558, 2010.

OLIVEIRA, L.B.; ACCIOLY, A.M.A.; MENEZES, R.S.C.; ALVES, R.N.; BARBOSA, F.S.; SANTOS, C.L.R. Parameters indicating the nitrogen mineralization potential of organic compounds. IDESIA (Chile), v.30, p.65-73, 2012.

PAUL, E.A.; CLARK, F.E. Soil microbiology and biochemitry. California: Academic Press, 1996, 340p.

PINTO, R.J.B. Namocomposites of cellulose and metals (Au and Ag). 2008. 84p Dissertation presented to the chemistry department of the University of Aveio, 2008.

REDIN, M. Biochemical composition and decomposition of aerial parts and roots of commercial crops and ground cover plants. 2010. 141p. Dissertation Federal University of Santa Maria, 2010.

REIS, T.C.; RODELLA, A.A. Kinetics of organic matter degradation and soil pH variation under different temperatures. Rev. Bras. Ci. Solo, v.26, p. 610-626, 2002.

RHEINHEIMER, D.S.; ANGHINONI, I.; KAMINSKI, J. Depletion of inorganic phosphorus from different fractions caused by successive extraction with resin in different soils and managements. Rev. Bras. Ciência do Solo, 24, p.345-354, 2000.

ROSCOE, R.; MERCANTE, M.F.; SALTON, J.C. Dynamics of soil organic matter in conservation systems: mathematical modeling and auxiliary methods. Dourados: Embrapa Agropecuária Oeste, p.165-199, 2006.

SANCHES, L.; VALENTINI, C.M.A.; BIUDES, M.S.; NOGUEIRA, J.S. Seasonal dynamics of litter production and decomposition in a tropical transition forest. Rev. Bras. Eng. Agric. Ambiental, v.13, p.183-189, 2009.

SILVA, I.R.; MENDONÇA, E.S. Soil organic matter. In: Novais, R.F.; Alverez, V.H.; Barros, N.F.; Fontes, R.L.F.; Cantarutti, R.B.; Neves, J.C.L. Fertilidade do Solo. Viçosa-MG, 2007, p.337-354.

SILVA, R.; HARAGUCHI, S.K.; MUNIZ, E.C.; RUBIRA, A.F. Applications of lignocellulosic fibers in polymer chemistry and composites. Química Nova, v.32, p.661-671, 2009.

SILVA, E.C.; MURAOKA, T.; BUZETTI, S.; ESPINAL, F.S.C.; TRIVELIN, P.C.O. Utilization of nitrogen from maize straw and green manures by the maize crop. Rev. Bras. Ci. Solo, v.32, p.2853-2861, 2008.

SOUTO, P.C.; SOUTO, J.S.; SANTOS, R.V.; BAKKE, I.A. Chemical characteristics of litter deposited in caatinga areas. Rev. Caatinga, 22, p.264-272, 2009.

TAIZ, L.; ZEIGER, E. Plant Physiology. Trad. de SANTARÉM, E.R. et al. 3 ed. PORTO Alegre, RS: ARTMED, 2004. 719p.

TAUK, S.M. Biodegradation of organic waste in soil. Rev. Bras. De Geociênia, 20, p.299-301, 1990.

TELES, M. M.; SANTOS, M. V. F.; DUBEUX JÚNIOR, J.C.B.; LIRA, M. A.; FERREIRA, R. L. C.; BEZERRA NETO, E.; FARIAS, I. Effect of fertilization and nematicide use on the chemical composition of fodder palm (Opuntia fícus indica Mill). Revista Brasileira de Zootecnia, v.33, n.6, p. 1992-1998, 2004.

TOSTO, M. S. L.; ARAÚJO, G. G. L.; OLIVEIRA, R. L.; BAGALDO, A. R.; DANTAS, F. R. Chemical composition and energy estimation of forage palm and dehydrated vine residue. Revista Brasileira de Saúde Prod. An., v.8, n.3, p.239-249, 2007.

VITTI, A.C.; TRIVELIN, P.C.O.; CANTARELLA, H.; FRANCO, H.C.J.; FARONI, C.E.; TRIVELIN, M.O.; TOVAJAR, J.G. Straw mineralization and sugarcane root growth related to planting nitrogen fertilization. Rev. Bras. Ci. Solo, v.32, p. 2757-2762, 2008.

ZHANG, D.; HUI, D.; LUO, Y.; ZHOU, G. Rates of litter decomposition in terrestrial ecosystems: global patterns and controlling factors. Journal of Plant Ecology, v.1, p.85-93, 2008.

CHAPTER 4

NITROGEN CYCLING IN A PASTURE SYSTEM

Erika Socorro Alves Graciano de VASCONCELOS;

PCI researcher at the National Semi-Arid Institute - INSA. Plant Production Area

Kaline Dantas TRAVASSOS;

PCI researcher at the National Semi-Arid Institute - INSA. Water Resources Area

Evaldo dos Santos FELIX;

PCI researcher at the National Semi-Arid Institute - INSA. Plant production area

Vanessa dos Santos GOMES;

Specialist in research methodology at the Paraíba State Research Support Foundation (FAPESq).

Daniela Batista da COSTA;

PCI researcher at the National Semi-Arid Institute - INSA. Plant production area.

INTRODUCTION

Nitrogen (N) cycling in cultivated pastures is a dynamic and complex process, considered to be one of the most important processes for the sustainability of pastures, above all because it is the most limiting element for production (Dubeux Jr. et al., 2004) and the main factor triggering the degradation process of pasture areas (Viana et al., 2011; Reis Jr. et al., 2004). The cycling of N in pastures cultivated in a tropical region, using an extensive system, without the use of nitrogen fertilizers, is represented by the connections between the soil, plant and animal compartments and the N transfer, input and output pathways, represented by the arrows in Figure 1.

In order to understand the cycling of N in cultivated pastures, this chapter presents a literature review describing the cycling of N in the soil, plant and animal compartments, carried out separately, addressing the main pathways of transfer, input and output of this nutrient and their respective estimates.

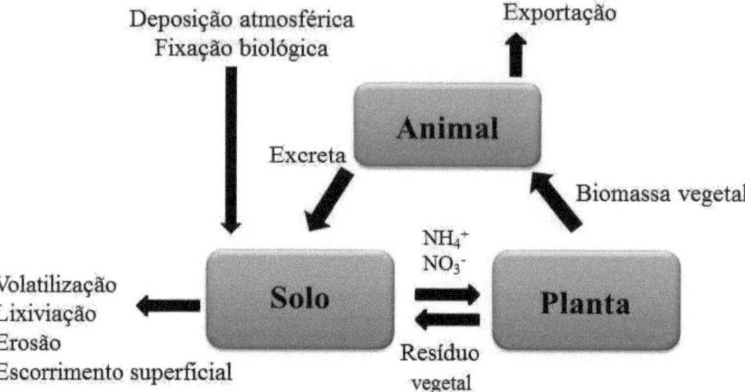

Figure 1. Nitrogen cycling in the main compartments of an extensively cultivated pasture system in a tropical region.

Nitrogen cycling in the soil compartment

The soil is an important compartment for N in pastures, playing a fundamental role in the reserve of this nutrient, both in inorganic and organic form. Approximately 98% of the total N in the soil is in organic matter in different molecules and with varying degrees of recalcitrance (Cantarella, 2007), and the smallest part is in inorganic forms composed of ammonium, nitrate and nitrite (NH$_4^+$, NO$_3^-$, NO$_2^-$) (Alfaia, 2006). Another proportion of N in the soil can occur in the form of N$_2$ and other NO$_x$ gases (Cantarella, 2007).

The atmosphere contains forms of combined N such as ammonium (NH$_4^+$) from industrial burning, volcanic activity and fires, and nitrate (NO$_3^-$) from the oxidation of N$_2$ by O$_2$ or ozone (O$_3$) in the presence of electrical discharges or ultraviolet radiation (Cordeiro, 2008). The way in which N enters the soil through atmospheric deposition is through rainfall, where the combined forms of N from the atmosphere are carried away by the water (Souza & Fernandes, 2006), known as wet deposition, mainly NO$_3^-$ and NH$_4^+$, and/or dust (Cantarella, 2007), coming from areas under erosion processes, through pollutants near industrial areas, or nitrate formed by the effect of lightning in storms, known as dry deposition, mainly N$_2$O and NH$_3$.

The amounts of N added by this process do not normally exceed 10 kg.ha^{-1}. For example, in different regions of São Paulo, Malavolta (1976) quantified contributions of only 10 to 15 kg ha^{-1}.year^{-1} of N. Araújo (2011), characterizing the total atmospheric deposition of inorganic N (NID = NO$_3^-$ + NH$_4^+$) and organic N (NOD) in the lower stretch of the Cachoeira

River, in the southern region of Bahia, reported an overall average deposition flux of 1.3 mol.ha^1.week^{-1} of NO_3^-, 2.7 mol.ha$_{(}^{-1})$.week$(^{-1})$ of NH_4^+ and 3.69 mol.ha$_{(}^{-1})$.week$(^{-1})$ of NOD. This contribution alone is insufficient to meet the demand of the production system, but it does serve to balance out denitrification losses in extensive grazing systems.

The way in which N enters the pasture system through biological nitrogen fixation (BNF) is carried out by free-living microorganisms associated with the plant species (Moreira et al., 2010) present in the soil, which have the enzymatic machinery needed to reduce diatomic N (N_2) to ammonia (Cordeiro, 2008). The availability of biologically fixed NH_4^+ in the soil occurs after the death of microbial cells and the lysis of cellular organic constituents (Reis et al., 2006) through the mineralization process.

In some pasture areas without the application of nitrogen fertilizers, BNF has maintained reasonable levels of productivity (Moreira et al., 2010). Braz et al. (2004) reported that BNF associated with Brachiaria can be a source of between 20 and 40 kg.ha^{-1}.year^{-1} of N. Silva et al. (2010), evaluating BNF in pastures with different cutting intensities in the Zona da Mata of Pernambuco, found a contribution of BNF by diazotrophic bacteria of around 10 to 42% in pastures made up of Brachiaria humidicola Rendle, B. decumbens Stapf. and Pennisetum purpureum Schum. These authors make the contribution of BNF in the soil very clear.

Organic waste deposited on the soil is the main source of N in organic form in the soil of pasture systems. This fraction is made up of animal waste and, mainly, plant waste at different stages of decomposition (Pegoraro et al., 2011; Selle, 2007), concentrated in the surface layers of the soil (0 to 20 cm) and refers to the deposition of animal excreta (urine and feces) and plant waste (Townsend, 2011). Mott & Popenoe (1977) report that up to 90% of nutrients, including N, can enter the soil through animal excreta. Most of the N present in urine is in the form of urea (70 to 90%), which is rapidly hydrolyzed by the enzyme urease to form ammonium (Haynes & Willian, 1993), which is why it is considered a readily available source of N in the soil (Wilkinson & Lowrey, 1973) and the rest consists of amino acids and peptides.

The input of N by animal waste is quite heterogeneous in the pasture (Euclides et al., 2007), as feces and urine are distributed unevenly. This return is influenced by the animal's stocking rate, grazing method, resting area, the animal (species, breed, sex), quantity and frequency of excretion, pasture management system, location of watering holes, topography of the land and shade (Rodrigues et al., 2008). Animals deposit their feces more in areas where they ruminate during the day or spend the night, while urine is excreted more in areas where they graze during the day (Rodigues et al., 2008). Depending on the management adopted, around 1 to 46% of the area is covered by feces, but the concentration of N in the region under the depositions of feces and urine is high, and can reach around 1,000 kg.ha^{-1} of N (Cantarella, 2007).

The deposition of plant residues (litter) in the soil compartment plays an important role in the recycling and conservation of this nutrient in the system. It is currently assumed that plant residue deposited on the soil is the main route responsible for incorporating N into tropical pastures (Schunke, 2001). In pastures, a significant amount of plant residue is deposited in the soil, as Cecato et al. (2001) found when assessing the accumulation of organic residue (known as litter) in an area of tanzania grass pasture managed at different heights under grazing with continuous stocking and variable load, they observed a large accumulation of litter, with an average of 2,179 kg.ha^{-1} of dry matter. This residue can be mineralized by soil microorganisms, transforming the N contained in the organic compounds into inorganic forms available in the soil (Correia & Andrade, 2008). Schunke et al. (2000), evaluating the growth and quality of *B. decumbens* in response to the cycling of N from the straw of this pure grass in soil subjected to three animal loads (0.6, 1 and 1.4 AU/ha), quantified the straw and N deposited on the soil and found that the treatments with pure brachiaria deposited 5,860 and 4,980 kg.ha^{-1} of straw and 69 and 45 kg.ha^{-1} of N, respectively.

N leaving the soil compartment in a cultivated pasture system can occur through volatilization, leaching, erosion, surface runoff and absorption by the plant root system. The release of N from the soil by volatilization occurs through the process of denitrification and the formation of ammonia from ammonium. Denitrification consists of the biochemical reduction of oxidized forms of N (NO_3^- and $NO_{(2)}^{(-)}$) to gaseous forms ($N_{(2)}$, N_2O and NO) (Victoria et al., 1992). These gaseous forms of N are volatilized into the atmosphere, thus leaving the soil compartment. The output of N by denitrification in pastures is high due to the high level of rapidly oxidizable carbon on the soil surface and the high concentration of NO_3^- present in the soil under the deposition site of feces and urine. The loss of NH_3^- by volatilization varies from 4 to 46% of the N contained in faeces and urine (Haynes & Williams, 1993), most of it through urine, because urea is easily hydrolysed to ammonia (Schunke, 2001).

The release of N by leaching occurs when this nutrient is carried away by the movement of water along the soil profile, mainly in the form of nitrate (NO_3^-), due to the low chemical interaction of this anion with soil minerals. This low interaction of NO_3^- with soil minerals is due to the predominance of negative charges in the soil, or at least in the surface layers of the soil (Cantarella, 2007), which means that its electrostatic adsorption is insignificant (Sagoi et al., 2003), and $NO_{(3)}^{(-)}$ is subject to leaching (Primavesi et al., 2006) into deeper layers of the soil, possibly reaching the water table. The estimated output of N by leaching varies from 8 to 20 kg.ha^{-1}.year^{-1}. Haynes & Williams (1993) reported N losses by leaching in intensively grazed pastures without the use of nitrogen fertilizers of 88 kg.ha^{-1}.year$^{(-1)}$.

The release of N through erosion and surface runoff takes place through a physical process of disintegration and transportation of soil particles, taking with it the organic and inorganic forms of N present. The disintegration of soil particles is mainly caused by the impact of raindrops directly on the soil surface, causing the aggregates to break down. In turn, the transport of particles and N occurs when the soil is saturated or when the infiltration capacity of water in the soil is lower than the rate of precipitation, causing excess water to run off the surface in sloping areas. This process depends on various factors, including vegetation cover and the slope of the area (Inácio et al., 2007). The outflow of N via these two routes in well-managed pastures is small and often does not exceed 5 kg.ha^{-1} of N (Costa et al., 2006).

Another way in which N leaves the soil is through absorption by the plant root system. In this process, the available inorganic forms of N, NH_4^+ and NO_3^-, present in the soil solution are absorbed by the plant roots through specific transporters (Reis et al., 2006), by active and passive mechanisms (with and without energy expenditure), and are subsequently translocated to the aerial part of the plants, forming part of organic compounds where they remain retained for a variable period of time.

Nitrogen cycling in the plant compartment

In the plant compartment, N is largely present in organic form, forming part of the structures of organic compounds such as amino acids, proteins and nucleic acid, and to a lesser extent in inorganic form (NH_4^+ and NO_3^-), which are used in the composition of organic compounds.

The way N enters the plant is through the absorption of NH_4^+ and $NO_{3}^{(-)}$ available in the soil, to meet the plant's physiological demand for this nutrient, a process called bioavailability. In this process, plant roots actively absorb nitrate from the soil solution through various low and high end nitrate-proton co-transporters. After absorption, nitrate is assimilated into organic compounds through its reduction to nitrite in the cytosol by the action of the enzyme nitrate reductase. The nitrite formed is rapidly transported into the chloroplasts in the leaves and the plastids in the roots, due to its high reactivity and toxicity. In these organelles, nitrite is reduced to ammonium (Taiz & Zeiger, 2009). The ammonium absorbed by the roots passively, i.e. without energy expenditure, is quickly assimilated into amino acids (Marenco & Lopes, 2005).

The ways in which N leaves the plant compartment are through the animal's consumption of plant biomass (extraction) and the deposition of senescent plant material. The output of N through animal consumption of forage biomass is due to the animal's need to meet the demand for nutrients from parts of the tissues in order to carry out its metabolic activities. The daily amount of forage consumed by the animal is a function of grazing time and ingestion rate (Gontijo Neto et al., 2006). According to Machado & Kichel (2004), the daily consumption of a 450 kg animal is approximately 11 kg of dry matter (DM). Therefore, in a brachiaria pasture with an N content of 0.83 % in the DM, the daily N extraction is 91.3 g per animal.

The deposition of senescent biomass from forage plants represents another outflow of N in the plant compartment. Although N is a nutrient that is easily translocated within the plant, via the xylem, in the transpiration stream, in the form of nitrate, amino acids (aspartate or glutamate), amides (asparagine or glutamine) or ureides (allantoin or allantoic acid) (Marenco & Lopes, 2005), significant amounts of N still remain in the senescent material, which is then deposited on the soil, forming litter. Xavier et al. (2011), evaluating the dynamics of litter in brachiaria pasture, quantified a monthly litter deposition of 623 kg.ha^{-1} of DM, thus estimating an annual litter deposition of 16,262 kg.ha$^{(-1)}$) of DM, and quantified the total N content of this material added to the soil at 107 kg.ha^{-1}.year$^{(-1)}$, confirming the return route of this nutrient to the plant compartment.

Nitrogen cycling in the animal compartment

The animal compartment is a circulating load on the soil compartment, which influences the transformation of N forms, the rate of cycling and availability, and is also an important component in pastures. In this compartment, N can be found as part of the composition of organic molecules in the animal's body, mainly proteins which are essential for the animal's development, and it can also be found in other forms present in the animal's digestive system.

The way N enters the animal compartment is through the consumption of plant biomass produced by forage, with a certain N content, to meet the animal's physiological needs to carry out its physical and metabolic activities. The amount of N that enters this compartment can vary depending on the type of forage consumed and the content of this nutrient in the plant biomass. Generally, the consumption of an animal weighing 450 kg live weight is approximately 11 kg of plant biomass per day, based on DM (Machado & Kichel,

2004), and the total N content in the dry plant biomass of brachiaria (hay) according to Fernandes et al. (2002) is 41.45%. It is therefore estimated that there is a daily intake of 4.5 g.kg^{-1} of N into the animal compartment through the consumption of plant biomass.

Only a small fraction of N is retained in the body mass. In the case of beef cattle, animals retain around 10% of the N consumed (Schunke, 2001) and this can vary according to the category, age, body and physiological condition of the animal, stage of production and level of forage consumption (Rodrigues et al., 2008). The amount of N that is not retained in the animal's body mass follows the pathways of exit from this compartment, which occurs through animal excretion (urine and feces).

In a B. decumbens pasture, Braz et al. (2002) evaluated the recycling of nutrients and the concentration of N in the feces of heifers with an average initial weight of 300 kg live weight, and estimated the daily output of N through the feces per animal at 22.1 g. They also reported that this output of N corresponds to 93% of the N consumed by the animal. According to Schunke (2001), beef cattle excrete around 90% of the N consumed through urine and feces, with the urine mainly in the form of urea or ammonia, and the feces mostly in organic form.

In addition to these ways of removing N from the animal compartment, the N that becomes part of the animal's body mass, in the form of proteins, can also be considered lost from the soil-plant system when the animal is removed from the pasture for slaughter. According to Boddey et al. (2000), this export of N rarely exceeds 10 Kg.ha^{-1}.year^{-1}.

FINAL CONSIDERATIONS

The efficiency of N cycling in pastures can contribute to greater bioavailability of this nutrient in the plant and animal compartments, which, as shown, benefits pasture production results. The deposition of organic residues, especially those of plant origin, represents the main way in which N enters the soil compartment to be made available to plants and animals, thus emphasizing its considerable importance in the pasture system. It is very important to point out that volatilization and leaching, among the ways in which N leaves the pasture, are of great concern, since a large amount of this nutrient is lost through these routes, which can compromise the entire system. Thus, there is a notable need to increase control over these N loss pathways in order to increase the efficiency of N cycling and, consequently, better production and greater pasture sustainability.

REFERENCES

ALFAIA, S.S. Characterization and distribution of organic nitrogen forms in three soils of central Amazonia. Acta Amazônica, v.36, p.135-140, 2006.

ARAÚJO, T.G. Total atmospheric deposition of nitrogen and major ions in the lower stretch of the Cachoeira River, Southern Bahia, BR. Dissertation (Master's degree in Tropical Aquatic Systems), 75f., Universidade Estadual de Santa Cruz, Ilhéus; BA, 2011.

BODDEY, R. M.; ALVES, B. J. R.; OLIVEIRA, O. C.; URQUIAGA, S. Pasture degradation and the nitrogen cycle. In: SALTON, J. C.; MELHORANÇA, A. L.; FONTES, C. Z. (Eds.). Workshop Nitrogen in the sustainability of intensive agricultural production systems. Dourados: EMBRAPA Agropecuária; EMBRAPA Agrobiologia, 2000. 163 p.

BRAZ, S.P.; NASCIMENTO JR., D.; CANTARUTTI, R.B. et al. Quantitative aspects of the nutrient recycling process by the feces of cattle grazing Brachiariadecumbens in the Zona da Mata of Minas Gerais. Revista Brasileira de Zootecnia, v.31, n.2, p.858-865, 2002.

BRAZ, S.P.; URQUIAGA, S.; ALVES, B.J.R.; BODDEY, R.M. Pasture degradation, soil organic matter and the recovery of productive potential in low technological input systems in the cerrado region. Seropédica: Embrapa Agrobiologia, 2004. 8p. (Embrapa Agrobiologia. Technical Circular, ISSN 1519-7328).

CANTARELLA, H. Nitrogen. In: NOVAIS, R.F.; ALVAREZ V., V.H.; BARROS, N.F.; FONTES, R.L.F.; CANTARUTTI, R.B.; NEVES, J.C.L. Fertilidade do Solo. Brazilian Society of Soil Science, p.376-449, 2007.

CECADO, V.; CASTRO, C.R.C.; CANTO, M.W.; PETERNELLIN, M.; ALMEIDA Jr., J.; JOBIM, C.C.; CANO, C.C.P. Forage losses in tanzania grass (PanicummaximumJacq cv. tanzania-1) managed under different grazing heights. Rev. Bras. Zootec., v.30, p.295-301, 2001.

CORDEIRO, L. Nitrogen fixation. In. KERBAUY, G.B. Fisiologia Vegetal, 2ed. Guanabara Koogan, p.51-64, 2008.

COSTA, K. A. P.; OLIVEIRA, I. P.; FAQUIN, V. Nitrogen fertilization for pastures of the Brachiaria genus in cerrado soils. Santo Antônio de Goiais: EMBRAPA arroz e feijão, 60p., 2006.

DUBEUX JUNIOR, J. C. B., SANTOS, H. Q., SOLLENBERGER, L. E. Nutrient cycling: Perspectives for increasing sustainability of intensively managed pastures. In: 21st Pasture Management Symposium. Proceedings of the 21st pasture management symposium. Piracicaba:FEALQ, p 357-400, 2004.

EUCLIDES, V.P.B.; COSTA, F.P.; MACEDO, M.C.C.; FLORES, R.; OLIVEIRA, M.P. Biological and economic efficiency of Tanzania grass pasture fertilized with nitrogen in late summer. Rev. Pesq. Agropec. Bras., v.45, p.1345-1355, 2007.

FERNANDES, L. O.; REIS, R. A.; RODRIGUES, L. R. A.; LEDIC, I. L.; MANZAN, R. J. Quality of brachiariadecumbensStapf. hay after treatment with ammonia amide or urea. Subjected to treatment with ammonia amide or urea. Rev. Bras. Zootec, v.31, p. 1325-1332, 2002.

GONTIJO NETO, M. M.; EUCLIDES, V. P. B.; NASCIMENTO Jr., D. N.; MIRANDA, L. f.; FONSECA, P. M.; OLIVEIRA, M. P. consumption and grazing time of tamarisk grass under different forage offerings. Rev. Bras. Zootec., v.35, p. 60-66. 2006.

HAYNER, R.J.; WILLIAMS, P.II.Nutrient cycling and soil fertility in the grazed pasture ecosystem.AdvancedAgronomy, v.49, p.119-199, 1993.

INÁCIO, E. S. B.; CANTALICE, J. R. B.; NACIF, P.G. S.; ARAUJO, Q. R.; BARRETO, A. C. Quantification of erosion in pastures with different slopes in the Ribeirão Salomea watershed. Rev. Bras. Eng. Agríc. Ambiental, V.11, p. 355-360, 2007.

MACHADO, L. A. Z.; KICHEL, A. N. Adjustment of scavenging in pasture management. Dourados: EMBRAPA agropecuária oeste; 55p. 2004.

MALAVOLTA, E. Manual de química agrícola - Nutrição de Plantas e Fertilidade do Solo. Editora Agronômica Ceres Ltda. São Paulo. 1976. 528p.

MOREIRA, F.M.S.; SILVA, K.; NÓBREGA, R.S.A.; CARVALHO, F. Associative diazotrophic bacteria: diversity, ecology and potential applications. Comunicado Scientiae, v.1, p.74-99, 2010.

MOTT, G.O.; POPENOE, H.L. Grasslands. In: ALVIM, P.T., KOZLOWSKI, T.T. (Ed.) Ecophysiology of tropical crops. New York: Academic Press, 1977. p.157-186.

PEGORARO, R.F.; SILVA, I.R.; NOVAIS, R.F.; BARROS, N.F.; FONSECA, S. Lignin-derived phenols, carbohydrates and amino acids in litter and soil cultivated with eucalyptus and pasture. Rev. Árvore, v.35, p.359-370, 2011.

PRIMAVESI, A.C.; PRIMAVESI, O.; CORRÊA, L.A.; SILVA, A.G.; CANTARELLA, H. Nutrients in the phytomass of Marandu grass as a function of nitrogen sources and doses. Ciência e Agrotecnologia, Lavras, v.30, n.3, p.562-568, 2006.

REIS, V.M.; OLIVEIRA, A.L.M.; BALDANI, V.L.D.; OLIVARES, F.L.; BALDANI, J.I. Symbiotic and associative biological nitrogen fixation. In: FENANDES, M.S. Nutrição Mineral de Plantas. SBCS, Viçosa-MG, p.154-172, 2006.

REIS JR., F.B.; SILVA, M.F.; TEIXEIRA, K.R.S.; URQUIAGA, S., REIS, V.M. Identification of Azospirillum amazonense isolates associated with Brachiaria spp. in different growing seasons and conditions and phytohormone production by the bacterium. Rev. Bras. Ci. Solo, v.28, p.103-113, 2004.

RODRIGUES, A. M.; CECATO, U.; FUKUMOTO, N. M.; GALBEIRO, S.; SANTOS, G. T.; BARBERO, L. M. Concentrations and amounts of macronutrients in the excretion of animals on mahogany grass pasture fertilized with phosphorus sources. Rev. Bras. Zootec., v.37, p. 990-997, 2008.

SAGOI, L.; ERNANI, P. R.; LICH, V. A.; RAMPAZZO, C. Nitrogen leaching affected by the form of urea application and management of oat crop remains in two soils with contrasting texture. Ciências Rural, V.33, p. 65-70, 2003.

SELLE, G.L. Nutrient cycling in forest ecosystems. Biosci. J., Uberlândia, v.23, p.29-39, 2007.

SCHUNKE, R. M.; RAZUK, R. B.; EUCLIDES, V. B. P. Production, decomposition and nitrogen release of litter from Brachiariadecumbens and Brachiariabrizantha pastures intercropped with Stylosanthesguianensis under two animal loads. In: REUNIÓN ALPA, 2000, Montevideo. Proceedings... Montevideo: ALPA, 2000. CD-ROM.

SCHUNKE, R. M. Pasture management alternatives for better use of soil nitrogen. Campo Grande: EMBRAPA gado de corte, 2001, 26p.

SILVA, L.L.G.G.; ALVES, G.C.; RIBEIRO, J.R.A.; URQUIAGA, S.; SOUTO, S.M.; FIGUEIREDO, M.V.B.; BURITY, H.A. Biological nitrogen fixation in pastures with different cutting intensities. Arch. Zootec., v.59, p.21-30, 2010.

SOUZA, S.R.; FERNANDES, M.S. Nitrogen. In: FERNANDES, M.S. Nutrição Mineiral de Plantas. SBCS, Viçosa, p.216-245, 2006.

TAIZ, L.; ZEIGER, E. Plant Physiology. Trad. de SANTARÉM, E.R. et al. 3 ed. PORTO Alegre, RS: ARTMED, 2009. 719p.

TOWNSEND, C.R. Nitrogen in pastoral systems. Porto Velho, RO: Embrapa Rondônia, 2011, p.29.

VIANA, M.C.M.; FREIRE, F.M.; FERREIRA, J.J.; MACÊDO, G.A.R.; CANTARUTTI, R.B.; MASCARENHAS, M.H.T. Nitrogen fertilization in the production and chemical composition of Brachiaria grass under rotational grazing. Rev. Bras. Zootec., v.40, p.1497-1503, 2011.

VICTORIA, R.L.; PICCOLO, M.C.; VARGAS, A.A.T. The nitrogen cycle. In: CARDOSO, E.J.B.N.; TSAI, S.M.; NEVES, M.C.P. Microbiologia do Solo. Campinas, p.105-119, 1992.

XAVIER, D. F.; LÉDO, F. J. S.; PACIULLO, D. S. C.; PIRES, M. F. A.; BODDEY, R. M. Leaf litter dynamics in Brachiaria pasture in silvopastoral and monoculture systems. Peq.Agropec.Bras.V.46, p. 1214-1219, 2011.

WILKINSON, S. R.; LOWREY, R. W. Cycling of mineral nutrients in pasture ecosystems. In: Butler, GW, Bailey, R.W (Eds). Chemistry and biochemistry of herbage.New York: Academic Press, 1973, V.2, p247-315.

CHAPTER 5

NUTRIENT LOSSES IN PASTURE AREAS

Erika Socorro Alves Graciano de VASCONCELOS;

PCI researcher at the National Semi-Arid Institute - INSA. Plant production area.

Vanessa dos Santos GOMES;

Specialist in research methodology at the Paraíba State Research Support Foundation (FAPESq).

Kaline Dantas TRAVASSOS;

PCI researcher at the National Semi-Arid Institute - INSA. Water Resources Area

Evaldo dos Santos FELIX;

PCI Researcher at the National Semi-Arid Institute - INSA. Plant production area.

Daniela Batista da COSTA;

PCI researcher at the National Semi-Arid Institute - INSA. Plant production area.

INTRODUCTION

Pastures are considered the basis of livestock farming and are one of the main sources of feed for beef and dairy cattle in many countries around the world. The forage species used for animal production stand out in importance as they occupy two thirds of the world's arable area (Borghi et al., 2018).

Brazilian agriculture has evolved over the last few decades. Currently, around 18.5% of the total area of the national territory is occupied by pasture (159 million hectares), 78% of which is under intermediate to severe degradation (Planalto, 2023), which hinders the ability to sustain animal productivity requirements.

The reduction in soil fertility due to nutrient losses (N, P, K, S, Ca and Mg) has been one of the main factors responsible for the degradation of pasture areas (Haddad & Alves, 2002; Peron & Evangelista, 2004), along with the lack of nutrient replacement (Pimenta et al., 2010). In this way, knowledge of the main routes of nutrient loss in pastures, especially the nutrients most required by plants, is of great importance for the proper management of pasture areas, and this knowledge can help in the development of conservation techniques that improve the use of nutrients in the system. This chapter presents a bibliographical review of the main ways in which nutrients are lost from cultivated pastures.

Nitrogen loss in cultivated pastures

Among the nutrients essential for plant growth and development, nitrogen (N) stands out as the most important in pasture systems (Dupas et al., 2010), as it is a constituent of proteins and directly interferes in the photosynthetic process as it is part of the chlorophyll molecule (Andrade et al., 2000). Losses of this nutrient from the soil to the atmosphere or to rivers, lakes and seas can occur through different processes, such as volatilization, denitrification and leaching (Cordeiro, 2008; Oliveira et al., 2007).

In tropical pastures, the volatilization of N in the form of ammonia (NH_3) is one of the main routes of N loss (Martha Jr. et al., 2004; Alves et al., 2011), which initially involves the hydrolysis of the nitrogenous compound by the enzyme urease, produced by bacteria, actinomycetes and soil fungi or even originating from plant remains (Costa et al, 2003).

The N that is added to the soil via organic animal waste, urine and faeces, which accounts for approximately 95% of the N ingested by animals, can be lost to a large extent through volatilization and leaching (Peron & Evangelista, 2004). Urine that manages to penetrate the soil can reduce nitrogen losses through volatilization and is an immediately available source for plants. Most of the N present in urine is in the form of urea (70 to 90%), which is rapidly hydrolyzed by the enzyme urease, forming ammonium (Haynes and Willian, 1993), as shown in the following equations, described by Almeida et al., (2008):

$$\begin{array}{c} H_2N \\ \diagdown \\ \quad C=O \\ \diagup \\ H_2N \end{array} \quad + \quad H_2O \quad \xrightarrow[\text{pH}]{\text{Urease}} \quad 2\,NH_3 \quad + \quad CO_2$$

Uréia — Amônia

$$NH_3 \quad + \quad H_2O \quad \longrightarrow \quad NH_4^+ \quad + \quad OH^-$$

Caráter Básico

In addition to the loss of N from organic animal waste, N from fertilizers used to fertilize pastures is also often lost from the system, largely through volatilization, as reported by some authors. Oliveira et al. (2007), evaluating nitrogen fertilization using urea on the efficiency of N recovery by the components of a *Brachiaria brizantha* cv. Marandu pasture system, found that N-urea losses occurred mainly through volatilization. Martha Jr. et al. (2004), estimating the loss of $N-NH_3$ by volatilization in Tanzania grass pasture fertilized with urea during a summer grazing cycle in Piracicada-SP, Brazil, found a high loss of ammonia by volatilization, corresponding to 44% of the N applied in the average of fertilizations with 40, 80 and 120 Kg/ha of N-urea.

According to Okumura and Mariano (2012), the loss of N through ammonia volatilization (NH_3) into the atmosphere is the main factor responsible for the low efficiency of urea fertilizer applied to the soil surface. This loss is intensified in systems where plant remains reduce the contact of urea with the soil, thus reducing the adsorption of NH_4^+ to organic and inorganic colloids, which further favors ammonia volatilization. As well as reducing the efficiency of nitrogen fertilizer use, the loss of N through volatilization reduces the profit of livestock enterprises based on the exploitation of fertilized pastures.

In situations of low oxygen availability in the soil, nitrate (NO_3^-) can be lost to the atmosphere by denitrification, in the form of nitrous oxide (N_2O), which is a greenhouse gas (Primavesi et al., 2006). Although this lack of oxygen in the soil is associated with waterlogged soils, denitrification problems can also occur in well-drained soils with compaction problems.

The NO_3^- available in the soil, when not absorbed by plants or immobilized by the soil microbiota, can be leached very easily because it has a negative charge and is not adsorbed by soil colloids which have predominantly negative charges (Primavesi et al., 2006). This process of N loss occurs through the descent of NO_3^- with the water in the soil profile to deep layers, where plant roots cannot reach, and can reach the water table, rivers and lakes.

Nitrogen losses through leaching in pastures, mainly in the form of NO_3^-, can reach up to 60% of N fertilizer application (Russelle, 1997). This loss can be reduced by using forage plants with deeper root systems, which can absorb this nutrient from deeper soil layers.

Losses of N through leaching are also considered to be very important, especially in pastures established on soils that are usually poor, sandy and have a low cation exchange capacity (CEC), which, combined with the large amounts of N applied to pastures in intensive systems (500 to 800 kg/ha/year), can also provide conditions of low ammonium retention, favoring losses through leaching (Oliveira et al., 2007).

The N that becomes part of the animal's body mass, in the form of proteins, is considered to be lost from the pasture system when the animal is removed from the pasture for slaughter. According to Boddey et al. (2000), this export of N rarely exceeds 10 Kg. ha^{-1}. year^{-1}.

According to Primaveri et al. (2006), controlling N losses in the pasture system is important because it represents losses of N from the soil available to plants; if lost in the form of N_2O, it could increase global warming and thus also reduce available water, as a result of greater evapotranspiration and more intense rainfall run-off; leached NO_3^- can enter rivers and groundwater and start the process of eutrophication in natural ecosystems.

Along with N, phosphorus (P) is of great importance in the renewal and maintenance of pastures with forage grasses in tropical conditions (Soares et al., 2000). Phosphorus is a crucial nutrient in plant metabolism, in the transfer of energy from the cell, respiration and photosynthesis, being a structural component of genes and chromosomes, as well as many coenzymes, phosphoproteins and phospholipids (Rezende et al., 2011), playing an important role in the development of the root system and in the tillering of grasses, its deficiency limits the productive capacity of pastures (Ieiri et al., 2010).

Phosphorus loss in cultivated pastures

P in the soil can be lost through the process of immobilization, where soil microorganisms immobilize P in microbial tissues in organic forms, and through the process of fixation, where P in inorganic forms is fixed to the surface of soil particles. The immobilization of P by microorganisms occurs when the organic residues in the soil have a C/P ratio greater than 300, which indicates insufficient P to meet the demand of the decomposing microorganisms (Moreira & Siqueira, 2006). P fixation, in turn, involves adsorption mechanisms, through electrostatic or covalent bonds, and precipitation, with the formation of insoluble compounds, making the element unavailable to plants (Corrêa et al., 2011). Pastures grown on acidic soils, with a clayey texture and, above all, high Fe and Al oxide content, may have low available P due to their high capacity for fixing it, as occurs in Latosols (Almeida et al., 2003).

Potassium loss in cultivated pastures

Potassium plays a fundamental role in plant nutrition, as it is the cation with the highest concentration in plants and is a nutrient with significant physiological and metabolic functions, such as enzyme activation, photosynthesis, assimilate translocation and also nitrogen absorption and protein synthesis, and is therefore a limiting nutrient (Andrade et al., 2000) in pasture systems.

Potassium (K) losses in pastures occur through runoff water or erosion and, above all, through leaching. According to Ferreira et al. (2009), significant losses of K through surface runoff can occur due to its presence in plant residues and in the topsoil. This is because K is not part of any organic molecular structure in plants, found as a free cation (Meurer, 2006), and can be easily removed by water after senescence.

Potassium in the soil solution is found in the form of an ion (K^+). As it is very mobile, it is easily leached in soils with a low CEC, such as sandy soils. According to Andrade et al. (2000), the loss of potassium in highly weathered, deep soils due to leaching is great and there is no significant accumulation.

Loss of Ca and Mg in cultivated pastures

Calcium and magnesium (Ca and Mg) also play important roles in pastures. Many of the functions of Ca in the plant are linked to the structural composition of macromolecules, located in the cell wall (apoplasm) and Mg is part of the composition of chlorophyll, directly related to the photosynthetic process, and also acts as an activator of enzymes related to the energy metabolism of plants, being a cofactor of almost all phospholative enzymes (Vitti et al., 2006).

The presence of Ca in the soil solution, in contact with the root system, is essential for plant survival, as this nutrient does not translocate from the aerial part to the new portions of the growing roots. Not only in the plant, but also in the soil, Ca has very limited mobility (Gebrim et al., 2008), unlike Mg, which, because it is not so strongly adsorbed to clay particles and organic matter, is subject to greater loss through leaching in the soil. The loss of Mg through leaching depends on the amount of water passing through the soil and its concentration in the soil solution, and can increase with the addition of soluble salts; fertilization with superphosphate and potassium salts (Vitti et al., 2006).

Sulphur loss in cultivated pastures

Sulphur (S) is a nutrient that plays an important role in the development of forage plants, as it participates in the formation of essential amino acids, such as cystine and methionine. Its deficiency interrupts protein synthesis, slowing down the growth of plants, even with an adequate supply of other nutrients, such as nitrogen (Oliveira et al., 2005).

The loss of S from the soil through extraction by plants can be high, as the S content in plant tissue is generally close to that of P, ranging from 2 to 5 g kg^{-1} of DM (Alvarez V. et al., 2007). In forage plants, S extraction occurs at around 50 kg ha^{-1} year$^{(-1)}$, considering productivity of 20 t ha^{-1}year$^{(-1)}$) of DM and S concentration in the aerial part of 2.5 g kg^{-1} (Oliveira et al., 2005).

Plants absorb S preferentially in the form of $SO_4^{(2-)}$ in soil solution, which has good mobility in the soil and can be lost through leaching (Alvarez V. et al., 2007), which is the main form of S-$SO_4^{(2-)}$ loss in the soil. The loss of $SO_4^{(2-)}$ by leaching is more intense in sandy soils than in clayey soils (Vitti et al., 2006).

The nutrients present in pastures can also be lost through erosion. According to Ferreira et al. (2009), nutrient losses by erosion can occur in the sediment carried from the soil and by runoff water. In this process of nutrient removal, organic matter and finer soil particles, both rich in nutrients, are more susceptible to loss, with organic matter being the first constituent to be removed due to its higher concentration on the soil surface and low density (Cassol et al., 2002).

When the soil is unable to meet the nutritional needs of forage plants, the photosynthetic capacity of the aerial part of the plant is compromised to such an extent that it is unable to meet its needs. At this stage, there is a drastic drop in production, the appearance of invasive plants with low palatability and the development of soil fauna that feed on the organic matter deposited, such as mound termites (Braz et al., 2004).

The evidence of the need to replenish nutrients in the pasture system, both for its recovery and to obtain good yields and management that allows for less nutrient loss in the system, is very clear. Several studies have frequently evaluated the effect of chemical fertilizers as a source of nutrients in the pasture system. Andrade et al. (2000) evaluated the effect of different doses of nitrogen and potassium on the dry matter yield of elephant grass and found that nitrogen and potassium fertilization had a positive influence on dry matter production.

Pellegrini et al. (2010), analyzing nitrogen fertilization in ryegrass pasture on the production and quality of forage mass, found that increasing the dose of nitrogen provides a higher rate of accumulation and production of total forage mass.

Observing the effect of increasing doses of nitrogen on the structural and agronomic characteristics of *Brachiaria brizantha* cv Xaraés under the climatic conditions of the southern region of the state of Mato Grosso, Cabral et al. (2012) found that nitrogen fertilization contributed positively to an increase in the number of total leaves of Xaraés grass and that increasing doses of nitrogen led to a greater tiller density of this forage, with the effect remaining throughout the year.

Ieiri et al. (2010), evaluating different phosphorus sources, doses and application methods, in terms of solubility, residual effect, availability in surface and incorporated applications, as well as production potential, during the recovery process of a degraded *Brachiaria decumbens* Stapf. pasture, observed that the application of increasing doses of phosphorus promoted increasing responses in dry matter productivity and the more soluble sources, such as triple superphosphate, showed higher phosphorus levels in plant tissues.

Gatiboni et al. (2000), evaluating the effect of different phosphates, soluble and natural, associated or not with liming, and the introduction of winter forage species in the improvement of natural pasture in sandy surface texture soil, observed that phosphate fertilization increased the dry matter productivity of the pasture, and soluble phosphates provide higher yields than natural phosphate. Rezende et al. (2011), evaluating the application of phosphorus in installments on the development of *Brachiaria brizantha* cv. Marandu, found an increase in the dry matter production of the aerial part and roots of the brachiaria when 100% of the P_2O_5 was used at sowing.

FINAL CONSIDERATIONS

Nutrient losses in pasture areas have led to degradation. There are several ways in which nutrients are lost in this system, with volatilization and leaching being the ones that need the most attention from livestock farmers, due to the great contribution these processes make to the loss of nutrients that are essential to the growth and development of forage plants.

The loss of nutrients from pasture areas, in addition to contributing to the degradation process, has led to the abandonment of these areas and encouraged the clearing of areas of native vegetation to establish new pastures.

REFERENCES

ALMEIDA, J.A.; TORRENT, J.; BARRÓN, V. Soil color, phosphorus forms and phosphate adsorption in latosols developed from basalt in the extreme south of Brazil. **Rev. Bras. Ci. Solo**, v.27, p.985-1002, 2003.

ALMEIDA, V.V.; BONAFÉ, E.G.; STEVANATO, F.B.; SOUZA, N.E.; VISENTAINER, J.E.L.; MATSUSHITA, M.; VISENTAINER, J.V. Catalyzing the hydrolysis of urea in urine. **Química Nova na Escola**, p.42-46, 2008.

ALVAREZ V., V.H.; ROSOOE, E.; KURIHARA, C.H.; PEREIRA, N.F. Sulfur. In: NOVAIS, R.F.; ALVAREZ V., V.H.; BARROS, N.F.; FONTES, R.L.F.; CANTARUTTI, R.B.; NEVES, J.C.L. **Fertilidade do Solo**. Brazilian Society of Soil Science, p.596-635, 2007.

ALVES, A.C.; OLIVEIRA, P.P.A.; HERLING, V.R.; TRIVELIN, P.C.O.; LUZ, P.H.C.; ALVES, T.C.; ROCHETTI, R.C.; BARIONI JR., W. New methods to quantify NH3 volatilization from fertilized surfasse soil with urea. **Rev. Bras. Ci. Solo**, v.35, p.133-140, 2011.

ANDRADE, A.C.; FONSECA, D.M.; GOMIDE, J.A.; ALVAREZ V., V.H.; MARTINS, C.E.; SOUZA, D.P.H. Productivity and Nutritive Value of Napier Elephant Grass under Increasing Doses of Nitrogen and Potassium. **Rev. bras. zootec.**, v.29, p.1589-1595, 2000.

ARAÚJO, E.A.; KER, J.C.; MENDONÇA, E.S.; SILVA, I.R.; OLIVEIRA, E.K. Impact of forest-pasture conversion on the stocks and dynamics of soil carbon and humic substances in the Amazon biome. **Acta Amazônica**, v.41, p.103-114, 2011.

BARCELLOS, A.O.; RAMOS, A.K.B.; VILELA, L.; MARTHA JUNIOR, G.B. Sustentabilidade da produção animal baseada em pastagens consorciadas e no emprego de leguminos exclusivos, na forma de banco de proteína, nos tropicos brasileiros. **Rev. Bras. Zootec.**, v.37, p. 51-67, 2008.

BODDEY, R. M.; ALVES, B. J. R.; OLIVEIRA, O. C.; URQUIAGA, S. Pasture degradation and the nitrogen cycle. In: SALTON, J. C.; MELHORANÇA, A. L.; FONTES, C. Z. (Eds.). **Workshop Nitrogen in the sustainability of intensive agricultural production systems** . Dourados: EMBRAPA Agropecuária; EMBRAPA Agrobiologia, 2000. 163 p.

BORGHI, E.; GONTIJO NETO, M. M.; RESENDE, R. M. S.; ZIMMER, A. H.; ALMEIDA, R. G. de; MACEDO, M. C. M. Recovery of degraded pastures. In: NOBRE, M. M.; OLIVEIRA, I. R. de (Ed.). Low-carbon agriculture: technologies and implementation strategies. Brasília, DF: Embrapa, 2018. chap. 4, p. 105-138.

BRAZ, S.P.; URQUIAGA, S.; ALVES, B.J.R.; BODDEY, R.M. Pasture degradation, soil organic matter and the recovery of productive potential in low technological input systems in the cerrado region. Seropédica: **Embrapa Agrobiologia**, 2004. 8p. (Embrapa Agrobiologia. Technical Circular, ISSN 1519-7328).

CABRAL, W.B.; SOUZA, A.L.; ALEXANDRINO, E.; TORAL, F.L.B.; SANTOS, J.N.; CARVALHO, M.V.P. Structural and agronomic characteristics ofBrachiaria *brizantha* cv. Xaraés submitted to doses of nitrogen. **Rev. Bras. Zootec.**, v.41, p.846-855, 2012.

CARDOSO, A.S. Evaluation of greenhouse gas emissions in different scenarios of intensification of pasture use in Central Brazil. Dissertation (Master's) - Federal Rural University of Rio de Janeiro, Postgraduate Course in Soil Science. 2012. 85p.

CASSOL, E.A.; LEVIEN, R.; ANGHINONI, I.; BADELUCCI, M.P. Nutrient losses by erosion in different methods of native pasture improvement in Rio Grande do Sul. **Rev. Bras. Ci. Solo**, v.26, p.705-712, 2002.

CARTER, M.S. Contribution of nitrification and denitrification to N2O emissions from urine patches. **Soil Biology and Biochemistry**, v.39, p.2091-2102, 2007.

CERRI, C. C.; BERNOUX, M.; MAIA, S. M. F.; CERRI, C. E. P.; JUNIOR, C. C.; FEILG, B. J.; FRAZÃO, L. A.; MELLO, F. F. De C.; GALDOS, M. V.; MOREIRA, C. S.; CARVALHO, J. L. N. Greenhouse gas mitigation options in Brazil for land-use change, livestock and agriculture. **Sci. Agric. Piracicaba**, v. 67, n. 1, p. 102-116, 2010.

CERRI, C.C.; MAIA, S.M.F.; GALDOS, M.V. et al. Brazilian greenhouse gas emissions: the importance of agriculture and livestock. **Scientia Agricola**, v.66, p.831-843, 2009.

CORDEIRO, L. Nitrogen fixation. In. KERBAUY, G.B. **Fisiologia Vegetal**, 2ed. Guanabara Koogan, p.51-64, 2008.

CORRÊA, R.M.; NASCIMENTO, C.W.A.; ROCHA, A.T. Phosphorus adsorption in ten soils from the state of Pernambuco and its relationship with physical and chemical parameters. *Acta Scientiarum. Agronomy*, v.33, p.153-159, 2011.

COSTA, N.L.; TOWNSEND, C.R.; MAGALHÃES, J.A.; PAULINO, V.T.; PEREIRA, R.G.A. Recovery and renovation of degraded pastures. **Revista Electrónica de Veterinaria**, v.7, p.9-49, 2006.

COSTA, M.C.G.; VITTI, G.C.; CANTARELLA, H. Volatilization of N-NH3 from nitrogen sources in sugarcane harvested without fire stripping. **Rev. Bras. Ci. Solo**, v.27, p.631-637, 2003.

D'ANDRÉIA, A.F.; SILVA, M.L.; CURI, N.; GUILHERME, L.R.G. Carbon and nitrogen stocks and mineral nitrogen forms in a soil submitted to different management systems. **Pesq. agropec. bras.**, v.39, p.179-186, 2004.

DIAS-FILHO, M.B. Os desafios da produção animal em pastagens na fronteira agrícola brasileira. **Rev. Bras. Zootec.**, v.40, p.243-252, 2011.

DUBEUX Jr., J.C.B.; MUIR, J.P.; SANTOS, M.V.F.; VENDRAMINI, J.M.B.; MELLO, A.C.L.; LIRA, M.A. Improving grassland productivity in the face of economic, social, and environmental challenges. **Rev. Zootec.**, v. 40, p.280-290, 2011.

DUPAS, E.; BUZETTI, S.; SARTO, A.L.; HERNANDEZ, F.B.T.; BERGAMASCHINE, A.F. Dry matter yield and nutritional value of Marandu grass under nitrogen fertilization and irrigation in cerrado in São Paulo. **Rev. Bras. Zootec.**, v.39, p.2598-2603, 2010.

EUCLIDES, V.P.B.; VALLE, C.B.; MACEDO, M.C.M.; ALMEIDA, R.G.; MONTAGNER, D.B.; BARBOSA, R.A. Brazilian scientific progress in pasture research during the first decade of XXI century. **R. Bras. Zootec.**, v.39, p.151-168, 2010.

FERNSIDE, P.M. Fogo e emissão de gases de efeito estufa dos ecossistemas florestais da Amazônia brasileira. **Estudos Avançados**, 16: 99-123, 2002.

FERREIRA, E.V.O.; ANGHINONI, I., CARVALHO, P.C.F.; COSTA, S.E.V.G.A.; CAO, E.G. Soil potassium concentration in a no-till crop-livestock integration system subjected to grazing intensities. **Rev. Bras. Ci. Solo**, v.33, p.1675-1684, 2009.

FIGUEIREDO, C.C.; RAMOS, M.L.G.; TOSTES, R. Physical properties and organic matter of a red latosol under management system and native cerrado. **Biosci. J.**; Uberlândia, v.24, n.3, p.24-30, 2008.

GALBALLY, I.E.; MEYER, M.C.P.; WANG, Y.P. et al. Nitrous oxide emissions from legume pasture and the influences of liming and urine addition. **Agriculture Ecosystems and Environment**, v.136, p.262-272, 2010.

GATIBONI, L.C.; KAMINSKI, j.; PELLEGRINI, J.B.R.; BRUNETTO, G.; SAGGIN, A.; FLORES, J.P.C. Influence of phosphate fertilization and the introduction of winter forage species on the forage supply of natural pasture. **Pesq. agropec. bras.**, v.35, p.1663-1668, 2000.

GEBRIM, F.O.; SILVA, I.R.; NOVAIS, R.F.; VERGÜTZ, L.; PROCÓPIO, L.C.; NUNES, T.N.; JESUS, G.L. Cation leaching favored by the presence of inorganic anions and low molecular mass organic acids in soils fertilized with poultry litter. **Rev. Bras. Ci. Solo**, v.32, p.2255-2267, 2008

HADDAD, C.M.; ALVES, F.V. Organic feed for cattle supplementation. In: **GLOBAL VIRTUAL CONFERENCE ON ORGANIC BOVINE PRODUCTION**, 1., 2002, Corumbá. http://www.

HAYNER, R.J.; WILLIAMS, P.II. Nutrient cycling and soil fertility in the grazed pasture ecosystem. **Advanced Agronomy**, v.49, p.119-199, 1993.

IEIRI, A.Y.; LANA, R.M.Q.; KORNDORFER, G.H.; PEREIRA, H.S. Sources, doses and modes of application of phosphorus in the recovery of pasture with *Brachiaria*. **Ciênc. Agrotec.**, v.34, p.1154-1160, 2010.

JUNIOR, C.C.; GOULART, R.S.; ALBERTINI, T.Z.; MAZZETTO, A.M.; FEIGL, B.J.; CERRI, C.E.P.; LANNA, D.P.; CERRI, C.C. Review: Greenhouse gas emissions and sustainability: A perspective on the beef production system in Brazil. **Rev. Bras. de Agropecuária Sustentável**, v.1, p.185-197, 2011.

LIEBIG, M.A.; GROSS, J.R.; KRONBERG, S.L.; PHILLIPS, R.L.; HANSON, J.D. Grazing Management Contributions to Net Global Warming Potential: A Long-term Evaluation in the Northern Great Plains. **Journal of Environmental Quality**, v.39, p.799-809, 2010.

KASCHUK, G.; ALBERTON, O.; HUNGRIA, M. Three decades of soil microbial biomass studies in Brazilian Ecosystems: Lessons learned about soil quality and indications for improving sustainability. **Soil Biology and Biochemistry**, Oxford, 42, 2010. 1-13.

MARTHA JUNIOR, G.B.; CORSI, M., TRIVELIN, P.C.O.; VILELA, L.; PINTO, T.L.F.; TEIXEIRA, G.M., MANZONI, C.S.; BARIONI, L.G. Ammonia loss by volatilization in Tanzania grass pasture fertilized with urea in summer. **Rev. Bras. Zootec.**, v.33, p.2240-2247, 2004.

MEURER, E.J. Potassium. In: FERNANDES, M.S. **Nutrição Mineiral de Plantas**. SBCS, Viçosa, p.281-298, 2006.

MOREIRA, F.M.S.; SIQUEIRA, J. O. **Microbiologia e Bioquímica do Solo**. 2 ed. Lavras: editora UFLA, 729p, 2006.

MUÑOZ, C.; PAULINO, L.; MONREAL, C.; ZAGAL, E. Greenhouse gas (CO_2 and N_2O) emissions from soils: A reviuw. **Chil. J. Agr. Res.**, v.70, p.485-497, 2010.

OENEMA, O.; VELTHOF, G.L,; YARNULK S. I.; JARVIA S.C. Nitrous oxide emissions from grazed grassland. **Soil Use and Management**, v.13, p.288-295, 1997.

OLIVEIRA, P.P.A.; TRIVELIN, P.C.O.; OLIVEIRA, W.S. Balance of nitrogen (^{15}N) from urea in the components of a marandu grass pasture under recovery at different liming times. **Rev. Bras. Zootec.**, v.36, p.1982-1989, 2007

OLIVEIRA, P.P.A.; BOARETTO, A.E.; TRIVELIN, P.C.O.; OLIVEIRA, W.S.; CORSI, M. Liming and fertilization to restore degraded *Brachiaria decumbens* pastures grown on an entisol. *Scientia Agricola*, v.60, n.1, p.125-131, 2003.

OLIVEIRA, P.P.A.; TRIVELIN, P.C.O.; OLIVEIRA, W.S.; CORSI, M. Fertilization with N and S in the Recovery of Pasture of *Brachiaria brizantha* cv. Marandu in Quartzarenic Neossol. **Rev. Bras. Zootec.**, v.34, p.1121-1129, 2005.

PLANALTO. Government issues decree allowing degraded pastures to be converted into plantation areas. Agenciagov, 2023. Available at: https://www.gov.br/planalto/pt-br/acompanhe-o-planalto/noticias/2023/12/governo-edita-decreto-que-permite-converter-pastagens-degradadas-em-areas-de-plantio. Accessed on; 13/12/2023.

PELLEGRINI, L.G.; MONTEIRO, A.L.G.; NEUMANN, M.; MORAES, A.; PELLEGRIN, A.C.R.S.; LUSTOSA, S.B.C. Production and quality of annual ryegrass submitted to nitrogen fertilization under grazing by lambs. **Rev. Bras. Zootec.**, v.39, p.1894-1904, 2010.

PERON, A.J.; EVANGELISTA, A.R. Pasture degradation in Cerrado regions. **Ciênc. Agrotec.**, v.28, p.655-661, 2004.

PRIMAVESI, O.; PRIMAVESI, A.C.; CORRÊA, L.A.; SILVA, A.G.; CANTARELLA, H. Nitrate leaching in coastcross pasture fertilized with nitrogen. **Rev. Bras. Zootec.,** v.35, p.683-690, 2006.

PRIMAVESI, O.; FRIGHETTO, R.T.S.; PEDREIRA, M.S. et al. Enteric methane of dairy cattle in Brazilian tropical conditions. **Pesquisa Agropecuária Brasileira**, v.39, p.277-283, 2004.

PIMENTA, L.M.M., ZONTA, E.; BRASIL, F.C.; ANJOS, L.H.C.; PEREIRA, M.G.; STAFANATO, J.B. Fertilidade do solo em pastagem cultivadas sob diferentes manejo, no nordeste do Rio de Janeiro. **Rev. Bras. Eng. Agric. Ambiental,** v.14, p.1136-1142, 2010.

REZENDE, A.V.; LIMA, J.F.; RABELO, C.H.S.; RABELO, F.H.S.; NOGUEIRA, D.A.; CARVALHO, M.; FARIAS JUNIOR, D.C.N.A.; BARBOSA, L.A. Morphophysiological characteristics of *Brachiaria brizantha* cv. Marandu in response to phosphate fertilization. **Rev. Agrarian**, v.4, p.335-343, 2011.

RUSSELLE, M.P. Nutrient cycling in pasture. In: INTERNATIONAL SYMPOSIUM ON ANIMAL PRODUCTION IN PASTURE, 1997, Viçosa, MG. Proceedings... Viçosa: Federal University of Viçosa, p.235-266,1997.

SANTOS, M.E.R.; FONSECA, D.M.; EUCLIDES, V.P.B.; RIBEIRO JÚNIOR, J.I.; NASCIMENTO JÚNIOR, D.; MOREIRA, L.M. Cattle production on deferred brachiaria grass pastures. **R. Bras. Zootec.**, v.38, n.4, p.635-642, 2009.

SILVA, T.G.F. Conceptual aspects, methods of analysis and information gathering for the livestock sector in Pernambuco in the face of climate change scenarios. In: GALVÍNCIO, J.D. **Mudanças climáticas e impactos ambientais**. Recife: Ed. Universitária da UFPE, p.25-64, 2010.

SILVA JR., R.S.; MOURA, M.A.L.; MEIXNER, F.X.; KORMANN, R.; LYRA, R.F.F.; NASCIMENTO FILHO, M.F. Estudo da concentração de CO_2 atmosférico em área de pastagem na região amazônica. **Rev. Bras. de Geofísica**, v.22, p.259-270, 2004.

SCHIERE, J. B.; IBRAHIM, M. N. M.; VAN KEULEN, H. The role of livestock for sustainability in mixed farming: criteria and scenario studies under varying resource allocation. **Agriculture, Ecosystems & Environments**, v. 90, p. 139-153, 2002.

SOARES, W.V.; LOBATO, E.; SOUSA, D.M.G.; REIN, T.A. Avaliação do fosforo de gafsa para recuperação de pastagem degradada em latossolo vermelho escuro. **Pesq. agropec. bras.**, v.35, p.819-825, 2000.

SPAROVEK, G.; CORRECHEL, V.; BARRETTO, A.G.O.P. The risk of erosion in brazilian cultivated pastures. **Sci. Agric. (Piracicaba, Braz.)**, v. 64, n.1, p.77-82, 2007.

TOWNSEND, C.R.; COSTA, N.L.; PEREIRA, R.G.A.Aspectos econômicos da recuperação de pastagens na Amazônia brasileira. **Amazônia: Ci. & Desenv.**, v.5, p.27-49, 2010.

VANLAUWE, B., AIHOU, K., HOUNGNANDAN, P., DIELS, J., SANGINGA, N., MERCKX, R., JENSEN, E. S., RECOUS, S. Nitrogen management in 'adequate' input maize-based agriculture in the derived savanna benchmark zone of Benin Republic. **Plant and Soil**, v. 228, p. 61-71, 2001.

VITTI, G.C.; LIMA, E.; CIGARONE, F. Calcium, Magnesium and Sulphur. In: FERNANDES, M.S. **Nutrição Mineiral de Plantas.** SBCS, Viçosa, p.300-322, 2006.

WILSEY, B. J. et al. Tropical pasture carbon cycling relationships betweem C source/sink strength, above-group biomass and grazing. **Ecology Letters**, v. 5, p. 367-376, 2002.

YDOYAGA, D.F.; LIRA, M.A.; SANTOS, M.V.F.; DUBEUX JÚNIOR, J.C.B.; SILVA, M.C.; SANTOS, V.F.; FERNANDES, A.P.M. Methods of recovering pastures of Branchiaria decumbens Stapf. in Agreste Pernambucano. **Rev. Bras. Zootc.**, v.35, n.3, p. 699-705, 2006.

yes
I want morebooks!

Buy your books fast and straightforward online - at one of world's fastest growing online book stores! Environmentally sound due to Print-on-Demand technologies.

Buy your books online at
www.morebooks.shop

Kaufen Sie Ihre Bücher schnell und unkompliziert online – auf einer der am schnellsten wachsenden Buchhandelsplattformen weltweit! Dank Print-On-Demand umwelt- und ressourcenschonend produziert.

Bücher schneller online kaufen
www.morebooks.shop

info@omniscriptum.com
www.omniscriptum.com

Printed by Books on Demand GmbH, Norderstedt / Germany